SEEDS OF CHANGE

J. EWING

SEEDS OF CHANGE

A History of the Ohio Agricultural Research and Development Center

text by Christopher M. Cumo
illustrations by Robert E. Whitmoyer

The Wooster Book Company
Wooster • Ohio
2000

The Wooster Book Company
205 West Liberty Street
Wooster Ohio • 44691
330-262-1688

Printed in the United States of America.

ISBN 1-888683-92-9

Cover: Charles E. Thorne, first full-time director of the Ohio Agricultural Experiment Station (now the Ohio Agricultural Research and Development Center) on horseback speaking to a group of farmers

Original line art by Jesse R. Ewing

Library of Congress Cataloging-in-Publication data

Cumo, Christopher
Seeds of change : a history of the Ohio Agricultural Research and Development Center / Christopher M. Cumo and Robert E. Whitmoyer
p. cm.
ISBN 1-888683-92-9 (alk. paper)

1. Ohio Agricultural Research and Development Center History.
2. Agriculture—Research—Ohio History.
I. Whitmoyer, R.E. (Robert E.)
II. Title.

S541.5.0320373 1999 99-32409
630'.7'20771—dc21 CIP

Printed on acid-free paper that meets industry standards for permanency.

CONTENTS

SEEDS OF CHANGE

PREFACE

In one way or another, we have been working on this history of the Ohio Agricultural Research and Development Center since 1991 (before 1965, the center was the Ohio Agricultural Experiment Station). We have persevered in our work because of our belief that farming and science are important activities. Indeed, we have tried to show that since its founding in 1882, OARDC has played a role in the world, not merely in Ohio and the United States. Whenever farmers have needed help they have been able to rely on the center. Among its achievements, its scientists have restored fertility to Ohio's soils, bred turkeys with legs strong enough to carry their bulky breasts, added beneficial supplements to livestock feed, and bred high-yielding corn, wheat, oats, soybeans, and tomatoes that are resistant to pathogens and insects. But farmers have not been the only winners. Consumers have gotten wholesome food at reasonable prices. Urban workers on their walk from car to office can pause in the shade of trees that were bred by OARDC scientists. Children drink vitamin-fortified milk pioneered by the center. Health-conscious Americans eat lean pork from a type of hog bred by the center. Homeowners fight Japanese beetles with the *Bag-a-Bug*, a product developed by U.S. Department of Agriculture scientists working at the center. 4-H clubs, the Cooperative Extension Service, and crop dusting are part of every rural county in America, and all have early ties to the center. Whether they know it or not, people of Ohio, the nation, and the world benefit from OARDC every day.

We also believe that history is important and that not enough has been written about OARDC's history. Its writing serves not only to recount the center's achievements, but to introduce OARDC to a new generation of Ohioans. In an era when Michael Jordan sells Nikes and McDonald's french fries, the center needs to trumpet its message of service to all Ohioans if it wants their continued allegiance. Andy Warhol's prediction that everyone would get fifteen minutes of fame may be enough for individuals but not for institutions. OARDC needs to live in the public eye. We hope this book is a step in that direction.

When we say that OARDC is an institution, we mean that it is people rather than buildings of concrete and sandstone. Among its scientists, William Krauss devoted half a century to the center, while Fritz Schmitthenner's career spans parts of two centuries and will probably surpass the tenure of Dr. Krauss. Charles Thorne and Edmund Secrest remained with the center, though as department chairs rather than directors, until their deaths. But for mandatory retirement, Roy Kottman would have remained director another decade. As it was, he stayed twenty-two years.

These people made our work satisfying but difficult. Large achievements and powerful personalities are not easy to recount without simplifying both. As historians, we had to be sensitive to science and to the psyche. Equally problematic, any institution with a long and proud tradition requires the historian to omit some people and events. Otherwise the story would never get told. We regret that space did not permit us to include everyone or to trace every scientific triumph in this story.

Enough for regret. This history is a celebration of more than a century of agricultural science in Ohio. In 1882, Governor Charles Foster predicted that the center would be no more useful than "a fifth wheel on a wagon." Was he ever wrong. May OARDC's next century be as fruitful as its first.

CHRISTOPHER CUMO is the author of **Seeds of Change: A History of the Ohio Agricultural Research and Development Center**. This book is based on and is an extension of his doctoral dissertation. His first book and several published articles have also been about OARDC. Chris writes for a variety of periodicals and is a contributing editor to the *adjunct advocate*. A native of Italy, he lives with his wife and two daughters in Canton, Ohio.

ROBERT W. WHITMOYER served the Ohio Agricultural Research and Development Center for more than thirty years as head of the Electron Microscope Laboratory (now the Molecular and Cellular Imaging Center). Since 1986 he has served as the center's historical records officer and since 1991 has been the curator of the OARDC Museum, which was created for the center's 1992 Centennial Celebration. Bob has restored several hundred vintage photographs and has organized the center's historical archives. Upon his retirement he has continued his position as historical records officer and has continued his work to preserve historical materials that relate to the center.

ACKNOWLEDGEMENTS

MANY PEOPLE GAVE THEIR TIME AND EFFORT IN MAKING THIS BOOK A REALITY. Kurt Knebusch, editor in the Section of Communications and Technology at the Ohio Agricultural Research and Development Center's Wooster campus, guided this book through editing and production, working closely with The Wooster Book Company. The OARDC librarian, Constance Britton, graciously allowed Chris Cumo unlimited access to published and archival documents. On the many occasions when he did not know where to turn for a bulletin or letter, she pointed him in the right direction. Roy M. Kottman, the director of the center between 1960 and 1982, allowed Chris to interview him at length. Dr. Kottman's vast knowledge and passionate support of agricultural science made him a remarkable man. It is regrettable that he did not live long enough to see this book in print. Among the scientists at the center, A.F. "Fritz" Schmitthenner, Lansing E. Williams, and William R. Findley are warmly thanked for allowing Chris to interview them. In addition, appreciation is expressed to Thomas L. Payne, the director of the center from 1993 through 1998, for his interest and support throughout this project.

In addition to conducting research and interviews at the center, Chris visited archives throughout Ohio and is especially indebted to two archivists for their patient assistance. On Ohio State's Columbus campus, Bertha L. Ihnat opened the university's archives to Chris and kindly assisted him in locating several documents. Of particular importance were the many letters of Alexis Cope, a secretary to the Board of Trustees, that Ms. Ihnet found. Likewise

Thomas A. Smith at the Rutherford B. Hayes Presidential Center helped locate several of Hayes' letters.

The illustrations in the book were provided by Bob Whitmoyer with assistance from Kenneth D. Chamberlain, publications photographer in the Section of Communications and Technology at the center's Wooster campus. They owe their vividness to David Wiesenberg, graphic designer and owner of The Wooster Book Company. He worked magic in restoring the large number of photographs from the nineteenth century that appear in this book. Beyond this work, he read the text with an eye to improving its clarity.

Finally, Chris and Bob thank their families—Chris's wife Gerianne and their daughters, Francesca and Alexandra; and Bob's wife Patty and their grown children Steven and Andrea—for their love and support.

1 IN THE BEGINNING

IN 1804 MERIWETHER LEWIS WROTE THAT THE MISSISSIPPI RIVER DRAINS "THE fairest portion of the earth." He could have said as much about the river's eastern tributary, the Ohio. The soils of its ancient floodplain yielded as many as 90 bushels of corn per acre during the 19th century, according to a farmer in Circleville, Ohio. The land nurtured the full range of crops that humans had domesticated during the past 10,000 years, including wheat, oats, potatoes, and tobacco. Pigs and cattle grew fat on thick stands of alfalfa.

That century began well for farmers. From the White House, Thomas Jefferson celebrated them as "the chosen people of God." A plantation owner, Jefferson vowed to preserve America as an agrarian republic. In it the virtues of self-reliance and honest labor were most evident on the farm. With these virtues, farmers would tame the vast prairie that stretched west with the promise of bountiful crops.

In Ohio, farmers could be as proud as anyone. In 1840 they produced 16.5 million bushels of wheat, more than any other state, and 33.7 million bushels of corn, second only to Kentucky. A decade later Ohio led the nation with 59 million bushels of corn. The bonanza farms, an editor of *Farm and Fireside* boasted, were not in the West, but right in the Buckeye State.

But as the century unfolded, Americans built railroads that charged farmers high rates and lowered food prices by speeding large surpluses to market. Industrialists forged an age of iron and then of steel as they built factories and cities and usurped the prestige that had once wreathed farmers. Perhaps worst of all, farmers exhausted soils in their frenzy to produce more food amid falling prices. The lament of worn-out soils arose in Europe and in the eastern United States during the 1830s and spread throughout the Midwest by the 1870s.

Ohio was no exception. Between 1850 and 1859 farmers harvested 12.2 bushels of wheat per acre but only 11.2 during the next decade. Between 1853 and 1868 yields dipped below 10 bushels per acre in six years, reaching a nadir of 4.5 in 1866. Corn fared better, dropping from 33.1 bushels per acre between 1850 and 1859 to 32.3 during the next decade. Although drought, insects, and frost contributed to these declines, in 1857 and again in 1860, John H. Klippart, the secretary of the Ohio Board of Agriculture, fingered soil exhaustion as the culprit. In 1880, Norton S. Townshend, an Ohio State University agricultural professor, observed that "even in Ohio, which 50 years ago was regarded as the most fertile of the new states, the evidence of partial exhaustion of the soil and the means for the restoration of its fertility have already become interesting topics for discussion."

Townshend and others believed that agricultural science would restore fertility to exhausted soils. In 1840, Justus Liebig, the celebrated German chemist, had announced that plants absorbed nitrogen through their roots in the form of ammonia and that farmers could increase crop yields by intelli-

Corn shocks at the Ohio Agricultural Experiment Station in Wooster. In the background at left is the Administration Building, which is still in use today. At right is the Bank Barn, which stood until 1964. The Apple Creek valley and the city of Wooster are in the distance. The photograph looks north (about 1909).

A typical Ohio farmstead at the turn of the 19th century. It is probably one of the station's branch farms.

gently adding ammoniacal fertilizers to their soils. Benjamin Silliman Jr., a chemist at Yale University and editor of the *American Journal of Science*, immediately praised Liebig's work as epochal, and in 1850, John P. Norton, a former pupil of Liebig, urged the state legislatures to create a system of agricultural colleges complete with experimental farms. Silliman and Evans Pugh, an agricultural chemist in Pennsylvania, echoed Norton during the 1850s and, when the states pled poverty, sought Congressional action.

Congress complied in 1862 by passing the Morrill Act, which granted each state 30,000 acres of public land per congressman to establish an agricultural and mechanical college or to maintain an extant one. The second option appealed to Michigan and New York, whose legislatures applied the grant to Michigan Agricultural College and Cornell University, respectively. Fifteen other states followed their lead while the rest, including Ohio, created land-grant colleges. Moreover, the act required these colleges to issue an annual report of their activities, including experiments, a provision they used as justification for establishing a farm.

These colleges were beholden to state legislators, however, who seemed less interested in agricultural improvement than in ensuring that their children received the liberal-arts training necessary to enter the urban professions. By the 1870s, this apparent emphasis angered farmers, who had initially hoped that these colleges would mold students into intelligent farmers. Instead, they felt that land-grant colleges had betrayed them by abandoning agricultural subjects in an attempt to mimic the prestigious New England universities. Consequently, farmers scoffed at the claims of administrators and faculty that land-grant colleges were the proper institutions to conduct agricultural experiments. They instead pointed to farms that fed students rather than conducted experiments, to meager enrollments in agricultural classes, and to the low esteem in which faculty and students generally held agricultural courses. Worse, few graduates of land-grant colleges returned to farming, leading farmers to regard these colleges as anti-farm.

The Ohio Agricultural and Mechanical College fared little better than its sister colleges during the 1870s. Although founded in 1870, it did not open until 1873 because the trustees could not define its mission. During this interlude, Townshend, then a trustee, and Joseph Sullivant, a proponent of liberal arts, wrangled over whether the college should emphasize practical subjects such as agriculture or the liberal arts. Sullivant's victory in 1873, and with it Townshend's departure from the trustees, enabled the trustees to institute a liberal-arts curriculum that included even ancient languages. Such a curriculum did not attract prospective farmers, and in 1877 only six of the 251 students were in agriculture. The next year the trustees renamed the college as The Ohio State University to reflect its broad mission to educate engineers, lawyers, and teachers.

This announcement incensed farmers, who at the annual Ohio Agricultural Convention in January 1879 focused their ire on Edward Orton, president of the university. When they cornered Orton on the subject of agricultural experimentation, he admitted that the university farm had performed few experiments. Later, Alexis Cope, the trustees' secretary between 1884 and 1904, conceded that farmers had, since the late 1870s, viewed the university's liberal-

An unidentified Ohio farmer in a tobacco field. The typical farmer of the 19th century needed help from agricultural research but at the same time was often suspicious of "book learning." The photograph dates from 1910–1920, possibly at Germantown, Ohio.

Station field workers cut and bind wheat by hand in this photograph taken in the 1930s. Their next step was threshing. The following spring they plowed under the stubble. This process remained the same from the founding of the station through the 1940s.

arts orientation and its lack of agricultural experimentation as betrayals of their interests. Thereafter, they frequently and loudly accused the university of abandoning them in order to become "the Harvard of the West."

As these criticisms suggest, the university was either unwilling or unable to ingratiate itself with farmers during the 1870s. It had only one agricultural professor, Townshend, and he was a physician rather than an agricultural scientist. In addition, it had no means of regularly communicating with farmers during the 1870s, since no Farmers' Institutes existed in Ohio until November 1880.

The most persistent attacks on the university came from Joseph Henry Brigham, master of the Ohio State Grange. During the Civil War, Brigham had served in the Ohio Regiment and had risen to the rank of colonel. Afterward, he returned briefly to the quiet life he treasured as a farmer, husband, and father of seven children. Upon joining the grange in 1873, however, Brigham, who stood nearly 6½ feet tall, quickly established a reputation as a forceful speaker and became master in 1878.

In this capacity, he temporarily elevated the grange from a social club into an advocate for agricultural experimentation, which he regarded as the tool of practical farmers. He believed that they could help themselves by farming more scientifically: for example, by adopting crop rotations and fertilizers that repeated trials had shown to yield the best. Although he enjoyed poking fun at

Workers load sugar beets into a horse-drawn wagon (circa 1910–1920). When the station was founded in 1882, horses, not tractors, pulled the plows and wagons.

politicians to score points with other farmers, Brigham had formed alliances with some of Ohio's most powerful men, including Rutherford B. Hayes and William McKinley, and tenaciously used his influence to support the development of agricultural experimentation in Ohio and nationwide. Though he would serve in the Ohio Senate in 1882 and run for the U.S. Senate in 1890, Brigham was first a farmer who believed that experimentation could only improve farming if it remained free from political considerations. He believed that Ohio State was little more than a political plum for the governor's cronies, and spoke for many farmers when he accused it of violating the Morrill Act by preparing students to become engineers rather than farmers.

The agricultural college was not, however, the only institution from which farmers could seek help. In England, where agricultural improvement had long been an avocation of the gentry, John B. Lawes, an eccentric and wealthy developer of patent medicines, had established a small greenhouse at Rothamstead in 1837. During the 1840s he expanded it into an experiment station complete with a barn for livestock feeding trials, experimental plots in which to compare crop varieties, crop rotations, and fertilizers, and laboratories for fertilizer analyses.

Meanwhile, Liebig's work had precipitated the rapid growth of a fertilizer industry throughout the German states. The need to ensure the quality of fertilizers prompted Saxony to create the first publicly funded station in 1851, which was originally little more than a laboratory for analyzing fertilizers. The German states established eleven stations by 1857, thirty by 1867, and seventy by 1875. By 1874 the stations were also conducting feeding and breeding trials with cattle, field experiments with fertilizers, experiments on tobacco and grapes, experiments on wine making and silk production, and tests of seed-germination rates. By then, it no longer made sense to distinguish between the Rothamstead station as primarily a farm for crop and livestock experiments and the German station as a chemical laboratory. Instead, by the 1870s these extremes had merged into a comprehensive model that combined experiments in the field and barn with chemical analyses.

As in the German states, the growth of a fertilizer industry in the United States exposed farmers to unscrupulous dealers. Because farmers generally had at best only a poor knowledge of chemistry and no way of analyzing fertilizers, they did not realize the extent of the fraud until the 1850s, when Samuel W. Johnson, professor of chemistry at the Yale Analytical Laboratory, began to publicize fertilizer analyses. Like several of his peers, Johnson had studied under Liebig at Giessen and had come to appreciate the value of the German stations as watchdogs of the fertilizer industry. With support from Wilbur O. Atwater, a chemistry professor at Wesleyan University, and the support of local farm organizations, Johnson successfully lobbied the Connecticut legislature in 1875 to create the nation's first agricultural

Unloading a wagon full of sugar beets in northwestern Ohio, around 1910–1920.

experiment station. As a fertilizer-regulatory agency, the Connecticut station reflected the German model of the early 1850s rather than the comprehensive one of the 1870s.

By 1880, state legislatures in North Carolina, California, and New Jersey had also established stations, none of which had the resources to conduct a broad range of experiments. Consequently, they, too, were fertilizer-regulatory agencies.

In addition to these publicly financed stations, Cornell University had a privately funded station that George Caldwell, professor of agricultural chemistry, had persuaded his colleagues and prominent farmers to form in 1879. He had expected state funds for his efforts, but the legislature demurred. The Cornell station instead subsisted on proceeds from food sales at the university farm until the university began to fund it in 1881. Like other stations, the Cornell station lacked the resources to undertake an ambitious agenda of experiments.

The difficulty of creating a station that had enough money to conduct experiments rather than merely analyzing fertilizers was evident in Ohio. The tussle between farmers and the university underscored their demand for an institution that would perform experiments. Mere fertilizer analysis could be lodged with any agricultural agency and did not warrant the creation of a new institution. The fact that farmers rejected the fertilizer-regulatory station as a model for an Ohio station, and with it the type of campaign Johnson had mounted in Connecticut, left unresolved the questions of how Ohio farm leaders would define a station and mobilize support for it.

John Klippart had tried to answer the first question as early as 1864, after he had returned from a tour of the stations in Möckern and Düsseldorf. At the annual meeting of the Board of Agriculture he asserted that Ohio Shorthorn breeders needed a station to perform the type of cattle-breeding experiments that had made German cattlemen the world's envy. Moreover, a station would, through experimentation, discover the principles of restoring Ohio's soil fertility. Furthermore, Klippart envisioned an experiment station as a natural component of a land-grant college since both would apply science to serve farmers. He therefore hoped that the Ohio legislature would create a land-grant college with an experiment station. Klippart repeated these arguments at the board's next meeting but apparently persuaded no one.

During the next thirteen years, neither Klippart nor any one else in Ohio advocated the creation of an experiment station. Perhaps Klippart contented himself that he had done his best to persuade others of a station's value. In any case, he thereafter turned his attention to an extensive agricultural survey of Ohio, the first part of which he finally presented in 1876. As for farmers, perhaps their hopes rested in the establishment and development of the Ohio Agricultural and Mechanical College rather than on an experiment station, at least until they became disenchanted in the late 1870s with its emphasis on liberal arts. They were doubtless well aware, as Brigham was, that the legislature had little enthusiasm for funding an experiment station.

Ultimately, farmers' criticism of The Ohio State University in 1879 may have prompted Edward Orton to resurrect the experiment station idea. At the Agricultural Convention that January, he attempted to excuse the fact that the university had performed few agricultural experiments by pleading insufficient manpower and resources. It could only conduct experiments, he claimed, if the legislature equipped it with an experiment station. Yet the fact that until then Orton had been silent about the value of a station suggests that he hoped to diffuse farmers' criticisms rather than conduct experiments. Indeed, Orton gave no indication that he had any conception of the scope of experiments that a station would undertake.

Orton's vague overture to farmers contrasted with Townshend's clear conception of the functions of a station. At the Ohio Shorthorn Breeders' Convention in January 1880, he argued that a station would, first, determine the best feeds for and breeds of cattle. Second, it would analyze fertilizers; identify the best crop varieties, rotations, and fertilizers for a particular soil and climate; and help farmers grow new crop varieties. Third, it would combat insect pests and diseases. These functions underscore that Townshend was promoting the comprehensive model of an experiment station that had developed in England and Germany by the 1870s. Although Townshend wanted a station because he believed it would benefit all Ohio farmers, he also apparently did so because he expected it to be subordinate to the university and he, as the agricultural professor and head of the university farm, would direct it. Despite his efforts, no evidence exists that Shorthorn breeders endorsed his proposal.

In his search for allies, Townshend had little choice but to sidestep Senator Brigham, who viewed the university

Joseph H. Brigham, who was a farmer, a colonel in the Civil War, a master of the Ohio State Grange, a state senator, and a member of the governing boards of both The Ohio State University and the Ohio Agricultural Experiment Station. In 1882 he guided through the Ohio legislature the bill that created the station.

farm as nothing more than a means for undergraduates to earn tuition by performing desultory and menial tasks and Townshend as a man incapable of understanding the importance of conducting experiments rather than merely talking about them. Although Brigham publicly voiced his distrust of Townshend and the university only in 1887, the bitter exchanges between the two on that occasion suggest that they were old enemies.

The tension between Brigham and Townshend was part of a national debate between systematic and scientific farming. Systematic farmers like Brigham saw farming as a business and themselves as producers. Experimentation could identify the most efficient farming methods, which experienced farmers would then tailor to their farms. As businessmen, they knew what was best for agriculture and should oversee experiments that scientists, who had the technical competence, would perform. Scientific farmers like Townshend retorted that farming was a profession rather than a business. Atop this profession stood scientists, whose experiments revealed how the laws of nature applied to farming. Because only scientists understood these laws, only they could oversee, design, and execute experiments. At most, farmers could apply the lessons from experiments that scientists had so painstakingly gained.

Norton S. Townshend was a physician, gentleman farmer, and reformer. A champion of agricultural science, he served as a trustee of the Ohio Agricultural and Mechanical College and was its first professor of agriculture. He believed the college needed an agricultural experiment station but was unable to get farmers to support the idea because they disliked the college's emphasis on liberal arts.

The rift between Brigham and Townshend underscored Townshend's need to seek allies elsewhere. He received an endorsement from *Ohio Farmer*, which stated on February 7, 1880, that Ohio needed a station in order to compete with German farmers who, thanks to their stations, had become more productive than their American counterparts. *Ohio Farmer* shared Townshend's vision of the comprehensive model and repeated his argument that a station would serve as a laboratory for soil and fertilizer analyses and germination tests of seeds, as a farm for comparisons of new and traditional crop varieties and of crop rotations and fertilizers, and as a barn for

A field demonstration of potato-treatment equipment about 1925.

selective breeding of livestock and comparisons of livestock rations. A station would also serve as a bureau of economic information by publicizing the varieties of crops and livestock that commanded the highest prices. Finally, an experiment station would provide practical experience in farm management for agricultural students at Ohio State.

At the same time, Townshend attempted to gain broad support from farmers. That February, *Ohio Farmer* reported that farmers who had attended the short courses at the university during the winter of 1879–1880 favored the creation of an experiment station as part of the university. Townshend, it seems, was using his position as the agricultural professor and the instructor of these courses to persuade farmers that an experiment station as part of the university would benefit them.

Meanwhile that month, Charles E. Thorne, then foreman of the university farm, advocated before the Cincinnati Grange the creation of an experiment station as part of the university. In addressing the grange, Thorne acted on Townshend's behest in order to gain its endorsement. In turn, William Hill, who was a member of that grange and a state representative, agreed to introduce Townshend's bill into the House with the grange's blessing. Thorne asserted that an experiment station would, like those in Connecticut and

North Carolina, protect farmers against fraudulent fertilizer and seed merchants. More than that, a station would help them introduce new crops and test livestock feeds, crop rotations, and fertilizers.

The Cincinnati Grange endorsed Thorne's position, and on March 9, Hill introduced Townshend's bill into the House. First, the bill made the proposed station part of Ohio State and the agricultural professor—in this case, Townshend—the director. It also made the trustees of the university the station's governing board. By subordinating the station to the university, Townshend hoped to overcome the shortage of manpower and resources that, as Orton claimed in 1879, prevented it from conducting experiments. Second, in accord with Townshend's broad vision of an experiment station, the bill called for a chemist, a botanist, an entomologist, a physiologist, and a microscopist to staff the station. Third, the bill requested $10,000, which was double the appropriation in 1880 for each of the stations in Connecticut, North Carolina, California, and New Jersey, "to erect buildings, purchase instruments and materials, and meet expenses for one year." Townshend justified this amount by arguing that a station could only serve Ohio farmers if it had scientists, land, buildings, farm implements, laboratory equipment, seeds, and livestock, none of which would be cheap. That the Hatch Act of 1887 would grant each station $15,000 indicates that Townshend's request was conservative rather than extravagant.

On March 13, Hill referred the bill to the House agricultural committee, which recommended its passage, unamended, three days later. However, the House never voted on it. Townshend learned too late that support from a handful of farmers did not mean that they had worked to pass his bill. Crestfallen, he met with other agricultural professors that June at Michigan Agricultural College to develop a plan for coordinating experiments at the land-grant colleges. The meeting produced no plan, only consensus that the colleges needed a network of stations if they hoped to be serious about experimentation. Townshend was adamant about this point and used the meeting to chide farmers for not being "awake to the importance of the measure" to create a station at Ohio State.

This explanation does not square with the fact that *Ohio Farmer* and farmers in Townshend's winter courses had favored his proposal. They, at least, had appreciated the bill's importance. Also, the fact that Hill introduced the bill suggests that, thanks to Thorne, the Cincinnati Grange had recognized its importance. Finally, Townshend surely would not have allowed Hill to introduce the bill had he believed farmers were unaware of its importance.

Nor is it credible, as Townshend and Cope later asserted, that Senator Brigham pressured House members to abandon the bill so he could later submit his own version and take credit for founding the station. Had this been the case, one suspects that he would have immediately introduced his own bill, which he did not. Rather, Brigham would not be drawn into the experiment station debate until early 1882. Brigham himself asserted that he had not known about Townshend's bill until after its debacle. Finally, this accusation does not square with Brigham's reputation for integrity. William Chamberlain would donate money to his campaign for the U.S. Senate in 1890 and would even spend three days campaigning for him because he believed Brigham sought to serve farmers rather than advance his own career. In 1890, Chamberlain described him as "a strong temperance man" and one who would never, in contrast to a widespread practice of the time, pocket money donated to his campaign. Also, Brigham's friendship with Hayes, who had a reputation for integrity, suggests his decency. That the trustees of Ohio State would appoint Brigham to their ranks in December 1887 suggests the esteem in which they held him. Throughout his career Brigham acted on the principle, without any apparent desire to advance himself, that agricultural experimentation would help farmers raise their living standards. Indeed, he would later complain privately that his hectic public life left him too little time to farm and to enjoy his wife and children.

Perhaps Townshend's bill failed because, as Brigham later observed, legislators were unwilling to create an experiment station if it meant an appropriation of $10,000. That they would create the Ohio Agricultural Experiment Station in 1882 with an appropriation of $3,000 suggests that Townshend had grossly misjudged their generosity. Equally important, he had not mobilized broad support from farmers for an experiment station in 1880, as he admitted. The fact that two years later farmers would enthusiastically support the creation of an independent station suggests that, although they probably appreciated the importance of a station, they were indifferent to one subordinate to the university. Aside from the *Ohio Farmer* endorsement, Townshend received no other public support; the Board of Agriculture, the Ohio State Grange, and the Horticultural Society were all silent on the issue of creating a station in 1880.

After the failure of Townshend's bill, farmers turned their efforts to fertilizer regulation. In January 1881, Hill introduced a bill into the House that would empower the Board of Agriculture to verify that fertilizers contained the kind and amount of ingredients stated on the manufacturer's label. The legislature passed the bill that May, guaranteeing that Ohio would not follow Connecticut's lead.

By then Connecticut had ceded leadership of the experiment station movement to New York, whose legislature in 1880 created a station with a broad

Horses pull an early harvester/thresher "combine" through an Ohio wheat field. Due to the size of the state's fields, the device actually saw little use here.

mandate to derive new knowledge about agriculture through experimentation. The onus of setting the station's agenda and location fell to a governing board whose president, Robert B. Swan, persuaded his colleagues in January 1882 to locate the station on land bordering his farm in Geneva. The creation of a true experiment station in New York signaled that farmers had abandoned the lure of a quick fix through fertilizer regulation.

In Ohio, William Rane Lazenby, who joined Ohio State in 1881 as a horticultural professor, followed New York's initiative. To this work Lazenby brought impressive credentials as a former professor of horticulture at Cornell University and former secretary of the Cornell station. His experience at Cornell, whose station was part of the College of Agriculture, had led him to view Ohio State as the natural home for a station. In January 1882 he asked the Board of Agriculture and the Horticultural Society to petition the legislature for a special appropriation to the departments of Horticulture, Botany, and Entomology. Although he did not mention the New York station as support for his overture, he was surely aware of it, and it may have prompted his request. Lazenby neither specified how much money he wanted nor precisely how he intended to spend it. Although he promised he would use it to teach agricultural subjects to farmers, he emphasized that it would enable him to conduct experiments. He did not outline the type of experiments he intended to undertake but promised "long, patient experimentation." This was the sort of talk that the Board of Agriculture and the Horticultural Society appreciated, since they were surely disappointed that Townshend had performed few experiments on the university farm. Indeed, Lazenby omitted Townshend from his proposal in an effort to distance himself from the latter's reputation for not conducting experiments. The fact that the Board of Agriculture and the Horticultural Society quickly endorsed Lazenby's proposal indicates their eagerness for an institution to conduct experiments.

At the same time, Lazenby sought Brigham's support, as was natural since the two shared a conviction that the application of science to farming, through experimentation, was the surest way to increase rural living standards as well as farmers' status. But Brigham opposed the creation of a station subordinate to the university, and the bill that Lazenby drafted with his help in early 1882 proposed an autonomous station. In contrast to Townshend's bill, it vested authority in an independent, five-member Board of Control rather than the university trustees. The governor would be an *ex-officio* member and the chairman. He would appoint three other members to one-year terms. The four would then elect a director, who would be the fifth member. Also in contrast to Townshend's bill, the Lazenby-Brigham bill did not specify the number and kind of staff and requested only $3,000, a concession to parsimonious legislators. Consequently, as Brigham anticipated, the station would be too poor to rent land and hire full-time staff. Such a station would be beholden to the university for land and scientists. The bill, therefore, would give the latter considerable influence over the station, which Lazenby wanted.

Although Brigham distrusted the university, he accepted a compromise and gathered endorsements from the Ohio State Grange, the Board of Agriculture, and the Horticultural Society that February. On March 1, he introduced Lazenby's bill into the Ohio Senate and the next day referred it to the Senate Agricultural Committee, which he chaired. On March 23, State Senator John D. Jones, surely with

William R. Lazenby, who wrote the bill creating the Ohio Agricultural Experiment Station. The Ohio legislature passed the bill in April 1882, and the governing board of the station rewarded Lazenby by appointing him the first director. But a lack of money and staff led him to resign in frustration in 1886.

Brigham's blessing, amended the bill to guarantee that the State Grange, the Board of Agriculture, and the Horticultural Society would each contribute one member to the Board of Control, and that all members "shall be practical farmers." That day, the Senate passed this version of the bill, and the House concurred on April 17. The virtual unanimity of support in both houses suggests that Ohio farmers were not anti-station, but rather anti-university. Opposition came predictably from Townshend, and also from Governor Charles Foster, who viewed an independent station "as useless as a fifth wheel on a wagon." Yet even he conceded that the "agricultural lobby" was too powerful to derail, and on April 17 he signed the bill creating the Ohio Agricultural Experiment Station "for the benefit of the interests of practical and scientific agriculture, and for the development of the vast agricultural resources of the State."

That July, a bitter Townshend criticized the legislature for not establishing the station subordinate to the university. As an independent institution, he asserted, the station would have neither adequate manpower nor resources to serve farmers. Its first five years would prove him right.

2 HARD TIMES

THE LEGISLATURE HAD CREATED A STATION ON PAPER. NOW THE BOARD OF Control needed to breathe life into it. As the first step in creating an institution that would echo with the sounds of human activity, the board met in Columbus on April 25, 1882, to appoint William Lazenby as director. His master of agriculture degree from Iowa State Agricultural College and his experience at Cornell legitimized the directorship and seemed to promise expert attention to farm problems. Without question, Norton Townshend had wanted the job, but his training as a physician and his failure to conduct experiments on the university farm disqualified him.

Lazenby now wore two hats as horticultural professor and station director, but the attempt to don both without knocking one aside was impossible. As professor, he continued to teach classes, grade papers, confer with students, and attend committee meetings. Such a schedule meant he could be only a part-time director, though the duties demanded much more from him. He needed to find the station a home, hire staff, set a research agenda, and spread the gospel of agricultural science to farmers.

This last task dogged him wherever he went. Farmers, of course, anticipated the dawning of a new era in agriculture, and in 1883 *Ohio Farmer* fed this expectation by predicting the station would help them coax an unprecedented bounty from their land and livestock.

And Lazenby wanted the reality of agricultural science in Ohio to match this rhetoric. He crisscrossed the state by train, meeting farmers at the State

The station's first building, located in Columbus at The Ohio State University. The photograph dates from 1885 and might be the oldest photograph of the station. The building was located near the present site of Townshend Hall.

Agricultural Convention, speaking at their institutes, and rubbing shoulders with them at the Ohio State Fair. He visited their farms to see damage from chinch bugs and to sample their whiskey. Whenever he met them he pumped their hands with animation and his mustache rose as his lips curved into a broad grin. Lazenby was at his best in these informal moments with farmers, his laughter joining in chorus with theirs. To be sure, he sometimes found himself deluged by their letters, but he always felt obliged to reply, and he never forgot that the station belonged to them.

Lazenby's schedule merely underscored the fact that his task would be more difficult than that of any subsequent director: to construct an experiment station in Ohio where none had existed, and to do so with insufficient money. As Lazenby noted in the first annual report, he had spent the rest of 1882 "organizing, equipping, and trying to make an experiment station, rather than directing work appropriate to one already established." The initial appropriation of $3,000 was at least $2,000 less than that of the other public stations in Connecticut, North Carolina, California, New Jersey, and New York. Emmett Mix, the Board of Control's treasurer, remarked in 1882 that this sum did not enable the station to serve farmers adequately. Indeed, the meager appropriation provided no salary in 1882 for the station's staff of William B. Alwood, its superintendent of field experiments; Nathaniel W. Lord, its chemist; and William S. Devol, its botanist. Lazenby himself received only $427.87, the balance that remained at the end of the fiscal year.

Furthermore, the small appropriation prompted the Board of Control to locate the station on Ohio State's Columbus campus, as Joseph Brigham had anticipated. The trustees provided the station with seventeen acres rent-free along the west bank of the Olentangy River, west of Neil Avenue. They even allowed the station to use the university farm's implements and two rooms in the chemistry building, as long as its staff paid for any chemical analyses. The fact that Lord and Devol were faculty at the university, and that Alwood was a graduate of the university as well as an employee, perhaps contributed to the trustees' benevolence. Yet as charitable as these terms were, they were part of a quid pro quo that required the station to give all its produce to the university except for what it required for seed.

Lazenby, of course, wanted the station to be much more than a cafeteria for the university, but poverty dogged his efforts. Although he admired the broad research of the German experiment stations that were created during the 1850s, Lazenby did not have enough money to conduct even a fraction of their research. Similarly, he could not hope to emulate the achievements of the station in Rothamstead, England, which since 1837 had thrived on a private endowment and did not need to court farmers and legislators, a chore that absorbed much of Lazenby's time and energy. Moreover, he could not demonstrate the station's utility to farmers by concentrating on fertilizer analyses, as the stations in Connecticut, North Carolina, and New Jersey did, because the Board of Agriculture had, since 1881, analyzed fertilizers in Ohio.

Seeking to establish a suitable role for the station, Lazenby distributed questionnaires in 1882, 1883, and 1884 to farmers throughout Ohio to learn what they viewed as their most intractable problems as well as their most important crops. These questionnaires also served to introduce the station to

A view of the station's seventeen acres in Columbus in 1887. In the center is the station's first building.

Test plots started by William R. Lazenby, the station's first director, in Columbus in 1883. The photograph was taken about 1885. In the background are buildings of the Ohio State campus.

farmers and to remind them that it existed to serve them. Yet Lazenby had difficulty defining this service more clearly than a general commitment to improve the raising of grains, livestock, horticultural products, and forest trees. Although no evidence exists concerning the number of responses to his questionnaires or the content of these responses, by 1883 Lazenby had identified the raising of corn and wheat, and particularly, determining the best fertilizers, as the greatest need of Ohio farmers. His comparisons of the effects of different fertilizers on crops, however, were poorly conceived. In 1882, for example, Lazenby applied manure, wood ashes, and coal ashes to plots of corn. Manure was a traditional soil amendment, wood ashes made sense given that slash-and-burn methods improved soil fertility, and coal ashes were rich in carbon, though he apparently did not appreciate their insolubility. He also applied calcium carbonate, commonly called lime, and plaster to other plots of corn, though he did not identify these soils as acidic. Finally, he added salt to a plot of corn, even though farmers had known since Roman antiquity that adding salt to the soil lowers crop yields.

Lazenby's experiments with fertilizers produced poor results. Between 1882 and 1885, he reported that fertilizers did not increase yields enough to cover the cost of fertilizers. Lazenby was apparently unaware that the alluvial soils of his experimental plots along the Olentangy River were so fertile they did not need fertilizers. He also apparently failed to realize that these soils did not typify most soils in Ohio.

That Lazenby's fertilizer comparisons failed did not deter him from other investigations. He compared yields of varieties of corn, wheat, oats, and potatoes despite the fact that Charles Thorne had conducted similar trials on the college farm in 1877, when he was foreman. Lazenby attempted to serve dairy farmers, who were then angry that unscrupulous farmers were selling margarine as butter, by analyzing products advertised as butter for the presence of vegetable oils. He also tested drinking water for contaminants, analyzed soils, and tested seed-germination rates. Although these were routine chemical analyses that any laboratory could perform, Lazenby, rather than shifting them to the Board of Agriculture, which already analyzed fertilizers, expressed eagerness to conduct them. Finally, the station collected weather data, advised homeowners on the care of rose bushes, shrubs, and turf grasses, and identified plant and insect species. Such a disparate agenda of projects underscores that Lazenby conceived of agricultural science in terms much more broadly, and diffusely, than experimentation.

So diffuse were the station's activities, and so meager its funds and size of staff, that Lazenby had to recruit farmers to conduct experiments in order to supplement staff and resources. For example, beginning in 1883 he recruited farmers to test varieties of corn and wheat on different soils. This was a common but questionable practice at stations that E. Lewis Sturtevant, director of the New York station, rightly condemned as unscientific. However, most of the early work at the stations concentrated on the necessity of gaining farmers' allegiance even when this meant sacrificing the quality of experiments.

Although Thorne privately criticized Lazenby, and Brigham told Rutherford B. Hayes, then president of the university's trustees, of his dissatisfaction with the station's course, Lazenby viewed a broadly defined approach as essential to popularizing the station. He recognized the need to publicize the station aggressively since, as he remarked in the annual report of 1883, "the station and its work are still imperfectly known to many citizens of the state." This view was natural for a man who measured success by the volume of letters he received from farmers. Always the publicist, he sent William Alwood to Athens, Ohio, in May 1884 in response to a letter he received three days earlier from a farmer whose apple orchard was suffering from canker worms. Alwood sprayed a mixture of kerosene, whale oil soap, and water on

Before the widespread use of the automobile, trains helped station scientists bring their knowledge to farmers. The photograph, possibly of the Hocking Valley Hog Train, dates from the 1910s. Today, sharing information with farmers—in person, in books, by the Internet—is still key.

the apples and advised the farmer to spray solutions of lead arsenate in the following days. The treatment worked, and Lazenby lost no opportunity to quote a letter of praise from the farmer in the annual report of 1885.

The need for farmers' support was perhaps more critical to the station during these formative years than at any subsequent time. Its appropriations of $3,000 in 1882 and $5,000 annually between 1883 and 1886 were lower than those of its counterparts in Connecticut, North Carolina, California, New Jersey, and New York during these years, and at best only half of the $10,000 that Townshend had set in 1880 as the minimum necessary to conduct agricultural research.

Nevertheless, the modest appropriation of $5,000 allowed the Board of Control to hire William J. Green as horticulturist in 1883 and to begin paying staff. The annual report of 1883 states that the Board of Control allocated $1,413.66 to salaries that year, though it does not specify how the sum was to be divided among Lazenby and his staff of four. In 1884, however, Lazenby's salary as director was $1,000. If this had been his salary in 1883, the four staff members would have shared $413.66, and since they held positions of equal status, each would have earned roughly $103 that year. With a total of $1,770 in salaries in 1884, each probably earned roughly $192, given Lazenby's $1,000 salary. As modest as these salaries were, one must remember that the staff worked part-time at the station. For example, Lazenby's salary as director was half his full-time salary of $2,000 in 1884 as professor of horticulture at Ohio State. The board also spent $1,357.82 on labor in 1883, $1,908.44 in 1884, and $1,973 in 1885. No evidence exists concerning how many laborers the station employed in any of these years or the hourly wage.

The station's modest growth between 1883 and 1885 gave it greater visibility. Lazenby, who remarked in 1883 that "the station and its work are still imperfectly known to many citizens of the state," was unaware that farmers were watching it, and he erred by not charting for the station a more independent course of action. He admitted that during 1883 he patterned comparisons of wheat varieties after those that Thorne had initiated on the college farm in 1877. According to Thorne, farmers became so irritated about the duplication of experiments that the trustees ordered Lazenby to appear at a special meeting on November 13, 1884. There the trustees accused Lazenby of misusing public funds by, in effect, plagiarizing the work of the university farm, and ordered him to return $750 to the state treasury. However, according to Alexis Cope, secretary of the university's trustees, the trustees ordered Lazenby to return the money only because they felt that his combined salary of $3,000 as director of the station and professor of horticulture at the university was excessive. Cope's version is unlikely, however, because the trustees could have quietly reduced Lazenby's salary had that been their objective. Instead, they ordered him to return $750 to the state treasury. In any case, the trustees could not compel him to return a portion of his salary as director of the station. A chastened Lazenby returned either $500 or $250, according to Thorne or Cope, respectively, and refused to accept further pay as director. Although he must have seethed privately, he publicly minimized this skirmish by stating in 1884 that "a spirit of harmony has existed between the station and all the departments with which it is so intimately connected." Even in a letter to his cousin in November 1884, Lazenby did not intimate his strained relationship with the trustees, but only remarked that "my work at the university continues much the same as usual."

Alexis Cope, one of a generation of Civil War veterans who distinguished themselves after Appomattox. A lawyer by trade, the former Union captain was secretary to the trustees of The Ohio State University and a confidant of Rutherford B. Hayes. Cope's letters reveal the tension between the university and the station between 1882 and 1892.

Thorne later argued that this dispute revealed the trustees' animus toward the station. According to him, they felt that it had usurped the university farm's prerogative to perform agricultural experiments. As late as 1927 Thorne would emphasize that the trustees would have preferred to plant grass on the university farm and at the station and rent the land to cattlemen for grazing. Although there may be some truth to Thorne's account, it does not square with the fact that they allowed the station to occupy seventeen acres and use farm implements, laboratory apparatus, and two rooms in the chemistry building, all free of charge.

That Lazenby duplicated some experiments of the university farm, and in the process angered farmers and the trustees, prodded the Board of Control to supervise the station more closely than it had. The 1884 annual report, which the board issued after the skirmish between Lazenby and the trustees, revealed that the board, for the first time, required the director to define and supervise all experiments. In addition, he had to summarize the station's progress at each board meeting and hold a staff meeting every Monday morning to learn the progress of experiments. This weekly meeting was to be attended by the superintendent of field experiments, the chemist, the botanist,

INDEPENDENT FARMERS' INSTITUTE

Schoenbrun, O., February 29, 1908.

PROGRAMME

MORNING SESSION—10:00 O'CLOCK

Invocation.............. ....Rev. Petty
Address of Welcome...................O. J. Demuth
Response............ ....C. F. Myer
Song by Grange..................................
Lecture—Smuts of Grain.......... .... Prof. A. D. Selby
Recitation..................... ......Miss Maud Watkins
Corn—Its Improvement............ .....C. F. Myer
Discussion..... ....Philip Bair
Duet...................Amanda and Bertha Schneider
Lecture—Corn Leaf or Blade Blight and Corn Mold
..............................Prof. A. D. Selby
Song by Grange................. ....

AFTERNOON SESSION—1:00 O'CLOCK

Music.. ....
Lecture—Bad Weeds and Impure Seeds..Prof. A. D. Selby
Recitation.............. ....Mrs. June Frey
Swine Growing For Profit............Harry S. Bartles
Discussion.............. ... .James Watkins
Recitation—"The Day the Hogs Go Off".........
....Miss Erma Williams
Lecture—Potato Blight and Potato Rot. Prof. A. D. Selby
Duties of a Farmer's Wife.........Mrs. James Watkins
Music.................................... ..

EVENING SESSION—7:00 O'CLOCK

Music.......... ....
Recitation.......... ....Miss Berth Kelly
Song................................ ....
The Farm, School and Home............ P. C. Knisely
Recitation............ .... Miss Erma Williams
Lecture—Notes of Travel............Prof. A. D. Selby
Recitation................. ....
Song by Grange.................. ....

Independent Farmers' Institute program, Schoenbrun, Ohio, 1908. Before the creation of the Cooperative Extension Service in 1914, Farmers' Institutes were a way for scientists to share their findings. The station botanist, Augustine D. Selby, lectured on five topics, from grain smut to potato rot. Note the recitations and music.

and the horticulturist, in addition to the director. The board required the superintendent of field experiments to supervise all laborers, keep data on his experiments, record the station's purchases and all sales of produce, and pay all bills. He also had to report all expenses to the director monthly, maintain the horses and implements, and collect weather data. The horticulturist supervised all experiments with fruits and vegetables, recorded his experiments, and identified all insects, in effect serving as the de facto entomologist. The botanist tested all seeds, identified plants sent to the station, and maintained a seed bank. In addition, he kept the mailing list and performed "such clerical and other work as [might] be assigned him from time to time." The chemist performed all analyses.

During 1886 the board transferred William J. Green's duties as entomologist to William Alwood and made William Devol the superintendent of field experiments. In turn, Green acquired Devol's former duties as botanist, meaning that instead of serving as horticulturist and de facto entomologist, he was now horticulturist and botanist. Finally, the board appointed H.J. Detmers as veterinarian. Through this shuffle, the board created the new positions of entomologist and veterinarian.

In November 1884, Lazenby complained that the station had grown too large for him to direct while still fulfilling his teaching duties at the university. To make matters worse, he knew that it was poorer than other stations and admitted to New York's E. Lewis Sturtevant in November 1885 that he was frustrated with its meager annual budget of $5,000. These complaints from a man who had stoically accepted a reduction in salary from the trustees suggest that the job was beginning to wear on Lazenby by the end of 1885. The strain of serving as director and horticultural professor apparently caused him to resign as director in January 1886 in order, he claimed, to devote himself to teaching at the university. His resignation left only Townshend to fill the vacancy and, in effect, unified the station and the university farm. According to Thorne, Townshend's ascendancy to the directorship that January placed the station "absolutely under the direction and control of the university."

Lazenby's accomplishments are difficult to assess. Under him, the station resembled a clearing house of information more than it did an experiment station. Yet this may have been necessary, as Lazenby realized, in order to gain broad support from farmers. Because the station had access to only seventeen acres in Columbus and operated with a maximum of five part-time scientists and a $5,000 annual budget, he needed the cooperation of farmers throughout Ohio to supplement staff and resources; and because the station did not pay well it could not attract and retain full-time scientists. Equally, if not more, detrimental to the quality of research was the nature of agricultural science before the Smith-Lever Act of 1914, which established an extension service at every land-grant college to teach farmers the lessons of agricultural science. The first agricultural scientists did not have the luxury of an extension service and therefore had to be both researchers and publicists to gain the support of their clientele, whereas most serious scientists wanted to confine their activities to research. Moreover, the professional scientist who admired the abstract qualities of the physical sciences usually disdained agricultural science for its practicality. Men attracted to the station had often, like Lazenby, grown up on farms and preferred the community of farmers and the open spaces of fields to the isolation of laboratories. Although the successful agricultural scientist had to be both researcher and showman, Lazenby erred by being too little of the researcher and left no one happy. Farmers did not get a scientific institution, Ohio State had a horticultural professor who could not devote himself full-time to teaching, and legislators could identify no tangible benefit from the station.

Two publicity shots by station photographers. At left, illustrating the difference between "short ear" and "tall ear" corn. The photograph dates from the 1920s. At right, a comparison of corn yields. The children are unidentified but were probably the children of station scientists. The photograph was taken in 1898.

3 FROM COLUMBUS TO WOOSTER

NO ONE HAD WATCHED THE STATION UNRAVEL AT THE SEAMS WITH GREATER ANXIETY than Charles Embree Thorne. He had been born in 1846 on a farm in Greene County, Ohio. From childhood the rhythm of planting and gathering crops was plowed into his soul. He was a Jeffersonian in believing that farmers were the bedrock of America and that farming was the noblest profession. Farmers were the creators of the nation's wealth and a repository of its virtues. America's center of gravity, he felt sure, was in the countryside rather than the city.

Thorne was nominally a Quaker, though agricultural improvement was the core of his faith. He studied agriculture for a year each at Antioch College and Michigan Agricultural College, but illness and the need to oversee the family farm led him to return home before completing a degree. Back in Ohio, he cut hay on the farm, wrote free-lance articles for agricultural journals, and was foreman of the university farm. Along the way he married Viola Brooke, whom he had met in East Lansing, Michigan.

Thorne's life was a busy but satisfying mix of tending to his wife and children and helping to edit *Farm and Fireside* when Townshend became director of the station in 1886. From his perch in Springfield, Thorne watched with growing anger as the university absorbed the station and its money and food, while returning nothing to farmers. He complained to Rutherford B. Hayes, then president of the trustees; to Joseph Brigham; and to anyone else who would listen. His typewriter thundered with indignation at the sight of the station crushed beneath the university's bulk.

But what could Thorne, Hayes, or Brigham do? The station had been vulnerable to a takeover because it lacked the money to assert its independence. Its autonomy on paper, Thorne scoffed, came with shackles in reality. And in 1887, the state legislature further impoverished the station by giving it only $1,500 in the belief that Congress would pour money into the coffers of every station, including Ohio's.

Indeed, during that decade, Washington, D.C., bustled with activity. Seaman A. Knapp, dean of agriculture at Iowa Agricultural College, led the charge for federal funding of the experiment stations. From Cornell University, Liberty Hyde Bailey, dean of the agricultural college, echoed Knapp's call for more money. And from the Tuskegee Institute, the young George Washington Carver joined Booker T. Washington in asking Alabama's senators and congressmen to vote money for agricultural science in general and for the stations in particular.

Charles E. Thorne, the first full-time director of the station, 1887–1921. His talents as a farmer, journalist, scientist, and administrator thrust him into the center of Ohio politics. His correspondents included George Washington Carver, Rutherford B. Hayes, William McKinley, and James M. Cox. *Ohio Farmer* called him "the grand old man of agriculture."

But if federal funds went to each state's land-grant college, as Knapp and Bailey hoped, Ohio State would grow rich but farmers would still get nothing. In an open letter to the Ohio legislature in March 1886, *Farm and Fireside's* publishers opposed the provision that would award federal funds to the university. They argued that the trustees were not farmers, but rather men hostile to agricultural experimentation who had shaped the university into a liberal-arts institution. That October, Thorne, in his role as associate editor of *Farm and Fireside*, amplified this criticism in an editorial by accusing the trustees of coveting Hatch funds with no intention of conducting agricultural experiments.

At the Ohio State Agricultural Convention on January 12, 1887, the tumult between farmers and the university reached a crescendo. There, William H. Scott, president of the university, asked the convention to adopt a resolution urging Congress to pass the Hatch bill. He and Townshend attempted to reassure their audience and also to defuse *Farm and Fireside's*

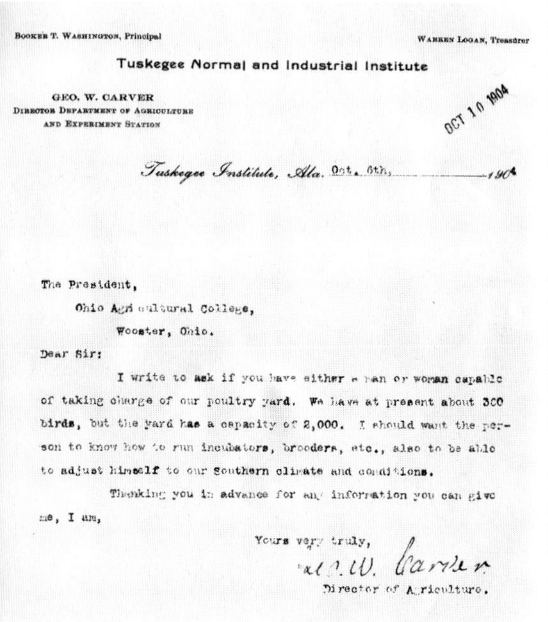

Booker T. Washington, Principal

Warren Logan, Treasurer

Tuskegee Normal and Industrial Institute

Geo. W. Carver
Director Department of Agriculture
and Experiment Station

OCT 10 1904

Tuskegee Institute, Ala. Oct. 6th, 1904

The President,
Ohio Agricultural College,
Wooster, Ohio.

Dear Sir:

I write to ask if you have either a man or woman capable of taking charge of our poultry yard. We have at present about 300 birds, but the yard has a capacity of 2,000. I should want the person to know how to run incubators, brooders, etc., also to be able to adjust himself to our Southern climate and conditions.

Thanking you in advance for any information you can give me, I am,

Yours very truly,

G. W. Carver
Director of Agriculture.

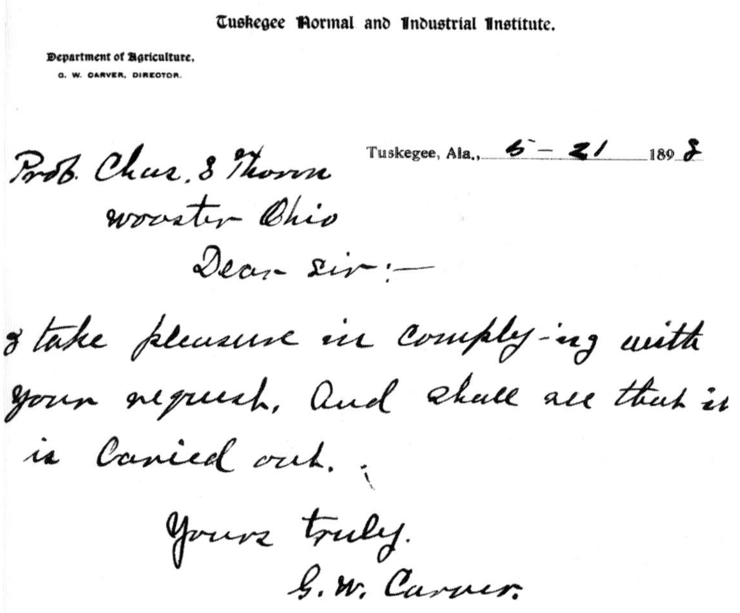

Tuskegee Normal and Industrial Institute.

Department of Agriculture,
G. W. Carver, Director.

Tuskegee, Ala., 5 - 21 1898

Prof. Chas. E. Thorne
Wooster Ohio

Dear sir:—

I take pleasure in comply-ing with your request. And shall see that it is carried out.

Yours truly.
G. W. Carver.

Two letters that Charles Thorne received from George Washington Carver, one in 1989, one in 1904. Carver was director of the Department of Agriculture and Experiment Station at the Tuskegee (Alabama) Normal and Industrial Institute. Booker T. Washington was principal of the institute.

attacks, by arguing that the Morrill Act had intended the land-grant colleges to conduct agricultural experiments and so had made Ohio State the proper recipient of federal money for agriculture—in this case, the Hatch funds. Whatever the merits of this argument, their position rested on the reality that the station and university farm were, in effect, unified under Townshend.

Their appeal did not persuade Brigham, who retorted that the university was little more than a plum for the governor's cronies who sought to advance their own careers rather than agriculture. Brigham was prepared to do more than talk. In his capacity as secretary of the National Grange, he intended to depart shortly for Washington. There, he planned to lobby Congress to amend the Hatch bill to grant federal funds to the governing boards of independent stations.

With these remarks, Brigham not only drew applause from the audience but touched a raw nerve at the university. Townshend, who was nursing a cold and trying to control a short temper, responded that if anyone sought his own advancement it was Brigham. Seth Ellis, who had been the first master of the Ohio State Grange, bluntly told Brigham that this amendment would defeat the bill. As a trustee of the university, Ellis may simply have wished to close ranks with Scott and Townshend. In any case, Ellis provided no evidence for his assertion even though Brigham pressed him for details.

These attacks led other members at the convention to defend Brigham, who needed no help. He silenced Townshend by asking him how many experiments he had conducted during the previous two years. At the end of this exchange, the convention endorsed Brigham's position by voting thirty-three to seven for the amendment.

Having won the battle in Ohio, Brigham and Thorne rushed to Washington, D.C., on January 20 in the hope that they could persuade Congress to pass a Hatch bill that would grant federal funds to the governing boards of independent stations. Hard on their heels, Norton Townshend and Lucius B. Wing, a banker from Newark, Ohio, and a trustee of the university, went to Washington to ask Congress to pass the Hatch bill unamended.

Brigham arrived in Washington to discover to his delight that on January 17, Massachusetts Senator Henry L. Dawes had introduced the amendment. The Massachusetts station, like its counterpart in Ohio, was separate from the land-grant college, and its farmers were equally adamant that their station, rather than the agricultural college, receive Hatch funds. Congress adopted it, and Grover Cleveland signed the Hatch Act on March 2, 1887. But due to congressional oversight, the act failed to authorize the treasury's comptroller to disburse the appropriation that year.

Consequently, the Ohio station had only $1,500 for 1887 because the state legislature had anticipated that the station would receive Hatch funds early

A view of the station in 1890. The photograph was taken from the tower of a building on the Ohio State campus in Columbus.

Charles Thorne, at right in the group in the background, speaks at Wheat Field Day in 1918.

that year. With less money than the station had ever had, Townshend could do nothing but suspend for one year the positions of botanist, entomologist, and veterinarian. With an irony that Thorne and Brigham must have appreciated, Townshend announced that the station would initiate no new experiments until the legislature gave him more money.

Despite this setback, Brigham and his colleagues on the Board of Control finally had the money they needed to free the station from the university's tentacles. As is the case with deadwood, Townshend had to go as director, and in his place they appointed Charles Thorne, though his editorial duties prevented his leaving Springfield until April 1888. Thorne arrived in Columbus full of energy and confidence and intent on pumping vitality into the station.

He was then 42 years old and had spent his life either farming or thinking about farming. In him, the board finally found a director who had the resolve and ability to mold the station into an institution that served Ohio's farmers. With remarkable clarity, Thorne perceived that in Ohio the improvement of soil fertility was then the primary task of agricultural science. In contrast to Lazenby's hit-or-miss approach, Thorne immediately began in 1888 to concentrate the station's work on restoring soil fertility. As a farmer himself, Thorne favored practical solutions to problems of production, and he possessed the ability to communicate these solutions to farmers in clear and direct language. He stressed that agricultural science required an inductive approach

At left, the Wooster town square around 1900. At right, an unidentified man and boy pause in front of a downtown Wooster dairy store.

and thereby accentuated experimentation as the primary function of an experiment station. Thorne, unlike Lazenby, embraced the trend in agricultural science toward professionalization, which began to transform the agricultural scientist from a publicist into an experimenter who sought to discover the "principles" of increasing food productivity.

Charles Thorne is on horseback in the center of this photograph taken at Wheat Field Day in 1918.

The trustees had no wish to see Thorne free the station from its doldrums and feared that a vigorous station might cause farmers to push the university down the path toward becoming a "cow college." To add to their irritation, the trustees were upset at Brigham for advocating an amendment to the Hatch bill that robbed them of federal funds. But they could not retaliate directly against Brigham and had to settle for plaguing Thorne, who they regarded as Brigham's lackey. In January 1888, the trustees began to charge the station $500 annual rent on the pretext that the Hatch Act enabled the station to pay its own way. Thorne countered that because the station more closely embodied the intent of the Morrill Act than did the university, the station had earned the right to a share in the land-grant heritage—in this case, the free occupancy of university land. But Thorne could not have it both ways; his insistence on the station's separate legal status gave the trustees every right to

This display shows the value of manure as a soil amendment. Until Charles Thorne conducted and publicized his research, many Ohio farmers considered manure to be useless. The photograph dates from 1900–1910.

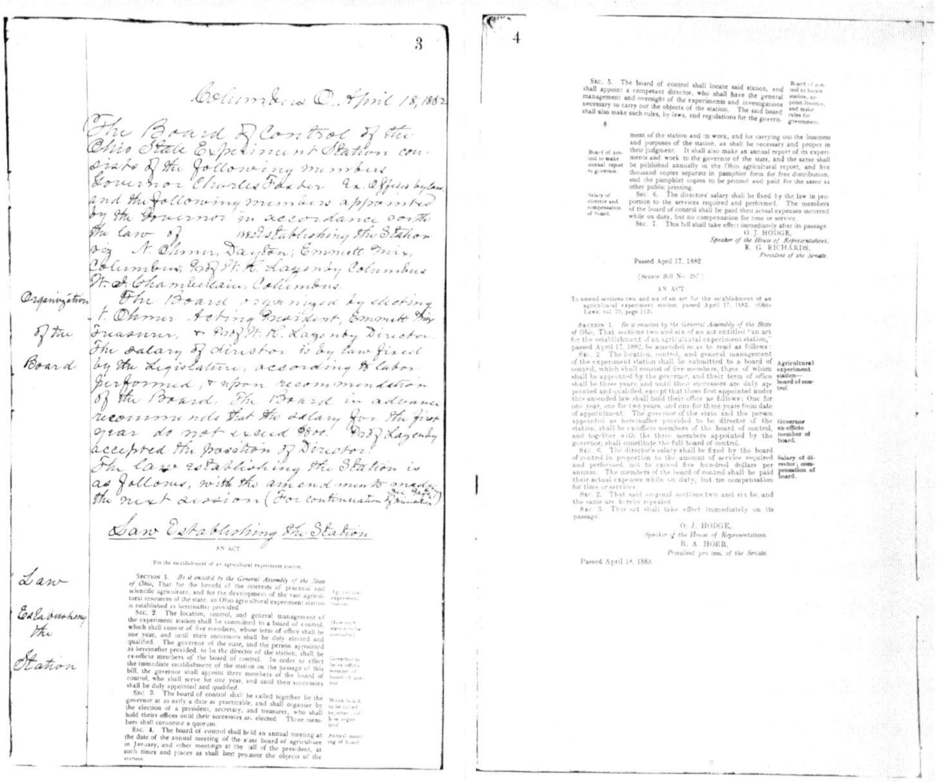

3

Columbus O. April 18, 1882

The Board of Control of the Ohio State Experiment Station consists of the following members Governor Charles Foster Ex Officio by law and the following members appointed by the Governor in accordance with the law of 1882 establishing the Station viz N. Ohmer, Dayton; Emmett Nix, Columbus; Prof W. R. Lazenby Columbus W. I. Chamberlain, Columbus.

Organization of the Board

The Board organized by electing N. Ohmer Acting President, Emmett Nix Treasurer, & Prof W. R. Lazenby Director. The salary of director is by law fixed by the Legislature, according to labor performed, & upon recommendation of the Board. The Board in advance recommends that the salary for the first year do not exceed $800. Prof Lazenby accepted the position of Director. The law establishing the Station is as follows, with the amendments made the next session (For continuation of minutes see page)

Law Establishing the Station

Law Establishing the Station

AN ACT

For the establishment of an agricultural experiment station.

SECTION 1. *Be it enacted by the General Assembly of the State of Ohio*, That for the benefit of the interests of practical and scientific agriculture, and for the development of the vast agricultural resources of the state, an Ohio agricultural experiment station is established as hereinafter provided.

SEC. 2. The location, control, and general management of the experiment station shall be committed to a board of control, which shall consist of five members, whose term of office shall be one year, and until their successors shall be duly elected and qualified. The governor of the state, and the person appointed as hereinafter provided, to be the director of the station, shall be ex-officio members of the board of control. In order to effect the immediate establishment of the station on the passage of this bill, the governor shall appoint three members of the board of control, who shall serve for one year, and until their successors shall be duly appointed and qualified.

SEC. 3. The board of control shall be called together by the governor at as early a date as practicable, and shall organize by the election of a president, secretary, and treasurer, who shall hold their offices until their successors are elected. Three members shall constitute a quorum.

SEC. 4. The board of control shall hold an annual meeting at the date of the annual meeting of the state board of agriculture in January, and other meetings at the call of the president, at such times and places as shall best promote the objects of the station.

4

SEC. 5. The board of control shall locate said station, and shall appoint a competent director, who shall have the general management and oversight of the experiments and investigations necessary to carry out the objects of the station. The said board shall also make such rules, by-laws, and regulations for the government of the station and its work, and for carrying out the business and purposes of the station, as shall be necessary and proper in their judgment. It shall also make an annual report of its experiments and work to the governor of the state, and the same shall be published annually in the Ohio agricultural report, and five thousand copies separate in pamphlet form for free distribution, and the pamphlet copies to be printed and paid for the same as other public printing.

SEC. 6. The directors' salary shall be fixed by the law in proportion to the services required and performed. The members of the board of control shall be paid their actual expenses incurred while on duty, but no compensation for time or service.

SEC. 7. This bill shall take effect immediately after its passage.

O. J. HODGE,
Speaker of the House of Representatives.
R. G. RICHARDS,
President of the Senate.

Passed April 17, 1882.

[Senate Bill No. 267.]

AN ACT

To amend sections two and six of an act for the establishment of an agricultural experiment station, passed April 17, 1882. (Ohio Laws, vol. 79, page 113.)

SECTION 1. *Be it enacted by the General Assembly of the State of Ohio*, That sections two and six of an act entitled "an act for the establishment of an agricultural experiment station," passed April 17, 1882, be amended so as to read as follows:

SEC. 2. The location, control, and general management of the experiment station shall be submitted to a board of control, which shall consist of five members, three of whom shall be appointed by the governor, and their term of office shall be three years, and until their successors are duly appointed and qualified, except that those first appointed under this amended law shall hold their office as follows: One for one year, one for two years, and one for three years from date of appointment. The governor of the state and the person appointed as hereinafter provided to be director of the station, shall be ex-officio members of the board of control, and together with the three members appointed by the governor, shall constitute the full board of control.

SEC. 6. The director's salary shall be fixed by the board of control in proportion to the amount of service required and performed, not to exceed five hundred dollars per annum. The members of the board of control shall be paid their actual expenses while on duty, but no compensation for time or services.

SEC. 2. That said original sections two and six be, and the same are hereby repealed.

SEC. 3. This act shall take effect immediately on its passage.

O. J. HODGE,
Speaker of the House of Representatives.
R. A. HORR,
President pro tem. of the Senate.

Passed April 18, 1883.

The law establishing the Ohio Agricultural Experiment Station and the organization of its Board of Control. It passed on April 17, 1882. O.J. Hodge, speaker of the Ohio House of Representatives, and R.G. Richards, president of the Ohio Senate, signed it. The document outlining the Board of Control was written the next day.

demand that it pay for the privilege of using university property. Although Thorne was sure the trustees had acted to anger him, he grimly paid the rent.

Despite his irritation, Thorne could do little short of moving the station. This idea did not immediately occur to him, as his own version of events implies, because in 1888 he used $3,000 from the Hatch fund to erect two buildings on campus for office and storage space. Moreover, Seth Ellis, who was then on the Board of Control, and his colleague Joseph Stevens had approached the trustees during late 1887 to improve relations between the two boards. In response, the trustees magnanimously elected Brigham to their ranks in December 1887. Though he had been severely criti-

After the station was moved to Wooster, it had only this building and a greenhouse to call home. The building was the first to be built by the station in Wooster (spring 1893). The Laboratory for Pest Control Application Technology occupies it today.

The station's Wooster campus in 1900. In the foreground are greenhouses that were new at the time; in the background, the first building (left) and the Administration Building.

cal of the university, Brigham thereafter marshalled his enormous energies toward improving relations between the station and the university.

Despite better relations, Thorne suggested at a joint meeting between the trustees and the Board of Control on January 13, 1890, that the station seek a new location. Columbus was unsatisfactory, he believed, because the alluvial soils near the Olentangy River were too rich for experiments with fertilizers and did not typify soils elsewhere in the state. The problems with the soils were a serious blow to Thorne's vision of an experiment station as an institution whose function lay in helping farmers throughout Ohio increase their yields through rotation and fertilization.

Thorne's argument did not then win a majority, and on November 28, 1890, the trustees and the Board of Control met again to discuss whether to move the station. Once more, Thorne did not win approval to seek a new location, and he now enlisted the support of Rutherford Hayes, then president of the trustees, in April 1891 and of farmers in April and May of that year. He argued that the station needed to move because the Olentangy's occasional flooding endangered spring plantings and because the city, which had recently laid sewer lines across land that the station rented, might soon wish to incorporate this property. These arguments persuaded Hayes, who, according to Thorne, believed as early as December 1887 that Columbus might soon incorporate the property on which the station was located. These arguments also persuaded the Board of Control, since Lucius Wing and Alexis Cope met with it on June 15, 1891, to attempt to convince its members to remain indefinitely in Columbus.

Thorne had no desire to remain, but only to reach detente with the trustees. Such a relationship would allow him, he hoped, to retain some portion of the university farm as the first in a network of experiment farms that

Seth H. Ellis, first master of the Ohio State Grange and a member of the governing boards of The Ohio State University and the Ohio Agricultural Experiment Station. He angered his colleagues on the Board of Trustees by favoring the station's move to Wooster.

he would use to tailor crop rotations and fertilizers to every soil type in Ohio. In order to mend fences with the trustees, Brigham, in February 1891, and Thorne the following April, separately approached Hayes to persuade him to improve relations between the other trustees and the Board of Control.

Thorne anticipated that Hayes would not encounter opposition from the trustees and on April 15 began soliciting bids for a new location from the counties. By July 28, he received offers of $75,000 each from Wayne and Clark counties and of $40,000 from Warren County. Each county, of course, also offered land. Thorne visited all three sites in late July and early August and by September 1, 1891, decided to accept Wayne County's offer, largely because its commissioners had sweetened their offer to $85,000.

But Thorne did not want to appear greedy, and so in the 1891 annual report he omitted any reference to money. He simply stated that he had favored Wayne County because its commissioners offered 470 acres outside of Wooster, the county seat, which afforded ready access to railroads and whose soils, which ranged from porous loams to moderate clays, typified large sections of eastern Ohio's glaciated soils. Wooster was also part of eastern Ohio's wheat belt. Unfortunately, Wooster lacked the central location of the university's campus, which distanced it from the State House and the headquarters of the Board of Agriculture and the Ohio State Grange. Yet Thorne viewed this as a minor sacrifice necessary to enable the station to begin its work to help farmers improve soil fertility.

The Wayne County commissioners proposed to raise the $85,000 through the sale of bonds, whose interest and principal would be paid by a property surtax. To gain voters' consent, the commissioners scheduled a levy for October 6, 1891. If successful, bonds would be available for purchase on December 17.

To win support for the October levy, Thorne addressed a gathering of farmers and local politicians on September 1 at a meeting of the Wayne County Farmers' Alliance. Here he adroitly introduced the issue of soil fertility by reminding his audience that wheat, the county's leading crop, had declined in yield per acre during the second half of the century. Moreover, the leading horticultural crop in Wayne County, the apple, had also declined in total production during this period. These declines exacerbated fears that already severe competition from western states might drive Wayne County apple growers from the market.

Thorne promised that if the station did no more than help the county's farmers increase wheat yields by one bushel per acre through intelligent rotation and fertilization—and here he hinted that their gains would be several times larger—they would increase their collective income by $20,000. He was less specific about the magnitude of increase that apple growers could expect from the station's work, though he assured them he would restore Ohio's importance as an apple-growing state.

The first water-pumping station for the city of Wooster, located at the north end of the city.

In addition to promises of higher yields, Thorne suggested that the station would enhance the county's status and increase land values. This was the kind of talk that Wayne County farmers wanted to hear; the alliance quickly endorsed the levy, the *Wayne County Democrat* followed suit on September 16, and voters assented on October 6.

Thorne had little opportunity to savor the victory, however, because Rezin B. Wasson, the Wayne County treasurer, attempted to block the move by suing the county commissioners sometime before December 17. He argued that the commissioners, in taxing Wayne County's residents to relocate an institution that benefitted all Ohioans and whose transfer should accordingly be borne equally by everyone in the state, would be violating the state constitution. At the same time, Wasson distributed a leaflet throughout Wayne County that outlined his ambition to create his own experimental farm to rival and ultimately ruin the station. The suit stunned Thorne, who considered Wasson insane. Whatever the merits of this charge, Wasson had tenacity and sufficient mental competence to pursue the case all the way to the Ohio Supreme Court. On July 6, 1892, the chief justice privately informed Thorne that the court would uphold the suit.

Thorne had no time to waste, and on the same day he got assurances from Governor William McKinley and the attorney general to proceed with the move to Wooster and to petition the state legislature to appropriate the $85,000 in lieu of the unconstitutional Wayne County levy. The next day Thorne hired a construction company to build a boiler house, an office building, and seven greenhouses at the Wooster site, and in early September he began to move equipment and seed there by wagon train. The romanticism of Thorne the pioneer blazing a new trail for the station is perhaps the

Horses and wagons from the station move to Wooster. Contrary to myth, much of the station's materiel went by train (1892).

most picturesque image in the station's history. On a more prosaic plane, however, he now faced the task that had overwhelmed Lazenby—to build a station were none existed.

To achieve this vision, the station needed money. In January 1893 at the State Agricultural Convention, Seth Ellis, then president of the Board of Control, announced that the board would petition the legislature to appropriate the $85,000. Ellis emphasized that the station existed to serve Ohio farmers, and that in the wake of Wasson's lawsuit farmers should rally behind the station and demand that the legislature release the money that rightfully belonged to it. The convention responded by voting unanimously to petition the legislature to appropriate the money for the move to Wooster. On April 25, 1894, the legislature finally gave the station the $85,000 that Thorne was sure Wasson was trying to steal from it.

Meanwhile, on January 11, 1893, Thorne asked the trustees to compensate the station $13,000 for the construction of the two buildings that the Board of Control had erected on campus in 1888. The trustees agreed only to discuss the matter further at a joint meeting on February 1. There they not only refused the board any compensation, but demanded that it pay the trustees "several hundred dollars" for the use of cows belonging to the university. At this point Thorne and Ellis stormed from the meeting, and negotiations were in danger of collapsing. Hayes' death on January 17 had left the trustees without its conciliator, as Lucius Wing admitted to Cope on January 20. Fortunately, Brigham and Chamberlain filled the void. Brigham pleaded with

Cope in a letter on February 20 to persuade the trustees to grant the board at least a little money, since the Wasson lawsuit "will sadly cripple us for a year or two." On March 11, Brigham wrote Cope again, stating only that he wanted "a fair and square settlement." Before mailing this letter, Brigham showed it to Chamberlain, who dashed off a postscript to Cope warning that the trustees and board should settle the dispute before it became public and tarnished their images. Their efforts apparently convinced the trustees, who on March 22, 1893, offered the board $5,000, an amount that allowed both parties to claim victory since the board received compensation and the trustees parted with less money than Thorne had wanted.

The compromise did more than settle accounts between the Board of Control and the trustees; it helped restore a working relationship between the two. For his part, Brigham performed his new role as trustee by quietly building consensus among the trustees that the university should emphasize mechanical training as well as the professions. For their part, the other trustees promoted agricultural experimentation, the subject dearest to Brigham and Thorne, by allowing the latter to conduct experiments on the university farm with consent from the professor of agriculture. After a difficult beginning, the station was finally going to establish itself as an institution independent of The Ohio State University.

4 CHARLES THORNE REBUILDS THE STATION

IN HIS UNPUBLISHED HISTORY OF THE STATION, CHARLES THORNE CELEBRATED THE Romans as antiquity's greatest farmers. The tribute was apt, for Thorne was a farmer and a builder in the Roman tradition. He arrived in Wooster with a reputation as a farmer, and once there, he did things in a hurry. This impatience was part of his personality. An anecdote has it that under Thorne's direction the station delivered milk to the town's residents. The delivery man was paid by the hour and therefore had reason to take the route at a leisurely pace. One day he was too ill to work, and Thorne couldn't find a replacement on short notice. In annoyance, he delivered the milk himself, and his irritation increased when he discovered that he completed the route forty-five minutes faster than the delivery man. The next day, Thorne asked the delivery man to explain how a man twice his age could deliver milk in less time. The man could give no answer, of course, and Thorne fired him at once.

But the desire to do everything right away also stemmed from Thorne's realization that farmers still did not have much of a station after ten years of paying taxes for one. This being so, Thorne set to work with an energy that had not been seen at the station before. Stone by stone, he crafted a renown as a builder to add to his considerable repute as a farmer. In the process, he amassed the sandstone architecture that remains the core of today's Wooster campus. In all, he erected the Administration Building; seven greenhouses; laboratories for biology, plant pathology, and entomology; a boiler house; a slaughter house; and buildings for dairy and beef cattle experiments.

Members of the Board of Control tour the station about 1915. The wagon is at the Administration Building.

The Administration Building was the centerpiece of this program, and it remains the most charming structure on campus. Upon its completion in 1897, Thorne invited to its dedication Governor Asa A. Bushnell; members of the legislature, including Joseph Brigham and William Chamberlain; and, of course, farmers. Ever the politician, Bushnell used the occasion to laud farmers for their productivity and the station as the nation's finest. Brigham, who was then assistant secretary of agriculture, summarized the importance of the day. It marked not the completion of a building but the continuation of what would be a long tradition of service to farmers. The station did not belong to him, Bushnell, Chamberlain, or Thorne. With his eyes squarely fixed on the farmers in the audience, Brigham proudly said that the station "is your institution."

The Administration Building stands today in silent tribute to Thorne. Its corridors once reverberated with his measured step and were pregnant with his ideas. Science was as dynamic in its rooms as in today's research laboratories. An inquisitive soul can still climb to the top floor, which leads to a steep staircase. These stairs make four right turns and open onto a small room near the top of the building's tower. From the center of this room another stairway ascends, this one a massive spiral of cast iron that ends at a four-foot-high door at the pinnacle of the tower. Only a three-foot-high railing protects a climber from a 100-foot plunge.

Two views of the same barn from the tower of the Administration Building. The photos date from the late 1890s and early 1900s.

This tower of iron and stone is not the Leaning Tower of Pisa, but all the same, one can imagine Galileo dropping iron weights from its height because of the strong imprint of history that it bears. From this summit, Thorne oversaw everything on the station's grounds. Others trudged here to collect weather data. A few may even have crept here on a dare. Whatever their reasons, for more than one hundred years this tower has elevated people above the earth and carried them nearer the heavens. The tower is alive with their desires, which form the current of history.

No less impressive than the Administration Building was the Bank Barn, which master craftsman John Starn of Smithville, Ohio, completed in 1894. It covered more than 1,000 square feet, stretching 230 feet long, 48 feet wide, and 60 feet high. The driveway by itself spanned as much area as an average four-bedroom house, sprawling 30 feet by 50 feet. Especially distinctive were the monogram OAES 1894 on one side of the slate roof and the words OHIO AGRICULTURAL EXPERIMENT STATION on the other.

The barn did not need to be so large to accommodate the station's cattle or match Thorne's ambition. It needed to be this size to allow room for experiments. The floor was at times cement, dirt, or drained gravel. The barn housed at least six milking stanchions, with dozens of milking machines.

Although in 1913 fire destroyed the creamery at the rear, the barn stood for 70 years and was torn down in 1964 only to clear land for the construction of Gerlaugh Hall. Even then, its timbers and wood-pinned joints were in perfect condition. Had Gerlaugh Hall been built elsewhere, who knows if the Bank Barn might still be standing in flawless splendor?

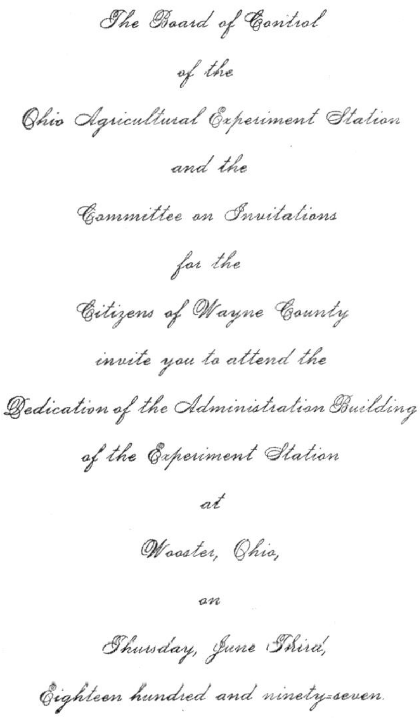
The Board of Control
of the
Ohio Agricultural Experiment Station
and the
Committee on Invitations
for the
Citizens of Wayne County
invite you to attend the
Dedication of the Administration Building
of the Experiment Station
at
Wooster, Ohio,
on
Thursday, June Third,
Eighteen hundred and ninety-seven.

The invitation to the dedication of the Administration Building, June 3, 1897. Farmers, legislators, and Governor Asa A. Bushnell attended.

These buildings gave Thorne a place to work, but where did he live? Without a car, he could not seclude himself in the countryside. And anyhow, that would not have been his style. He had to be in the thick of things, and so he stayed near town, where he sunk roots at the corner of Secrest Road and Madison Avenue. There he had a post-Victorian house built in 1894. Like every aspect of his life, Thorne supervised its construction from foundation to shingles. The exterior was red brick, and a patio in the style of Monticello led into the house. A photograph of his dining room shows a wooden table and chairs, china cabinet, chandelier, and picture on one wall. The furnishings in the rest of the house might have been as ornate, though one cannot know because additional photographs do not exist.

Tourists who want to see how Thorne lived can drive past the intersection of Secrest and Madison, for the house still stands. Although subsequent owners have refurbished the interior many times, the exterior is as it was in Thorne's day. To be sure, the house never rivaled the mansions of Henry Clay Frick and John D. Rockefeller in Cleveland, though it was grand by the standards of Wooster. It had to be; Thorne was not just the station's director, he was president of the Wayne Building and Loan Company. But he remained a grower at heart. Other directors, including Carlos Williams, moved into town so their children would be closer to schools.

Back at work, the station's buildings needed power, which they got through tunnels that snaked as far as 30 feet below ground and that originated at the boiler house. As early as 1893, steam boilers pumped heat into the first greenhouse and the biology building.

By 1897, the station included (top, left to right) the Administration Building; buildings for chemistry, entomology, and botany; and (bottom) the Bank Barn.

By 1920, electricity rivaled steam as a power source at the station, though at first the connection between electricity and steam power was intimate. After all, a steam engine powered the generator that first gave the station electricity in 110-volt direct current. City power soon supplanted the generator and, along with long-distance telephone service, linked the station to the modern world.

At the same time, workers at the station became more mobile as the Model T replaced the horse. Although Thorne, like Albert Einstein, never learned to drive, he authorized the first automobile purchase at the station. The arrival of cars at the station did not simplify life there, for Thorne had neither the tools nor mechanics to fix them. As late as 1920, he had Model Ts repaired at a garage in town. On one occasion, mechanics installed a new front fender, bumper, and headlight lens and adjusted the timing on a Model T for $12.50. Today a simple oil change costs twice as much.

Sawyers cut wood for use in building the Bank Barn (1893). The lumber came from station land in what is now Secrest Arboretum.

By horse and buggy and by Model T, scientists come to work in 1917. The photograph was taken from the Administration Building. In the background is the city of Wooster.

Model Ts crowd the grounds at the 1918 Wheat Field Day.

As everyone knew, the station was much more than stone, mortar, and Model Ts; it was a community of people bound together by the common purpose of serving farmers. The creation of this community owed much to Viola Thorne. When she arrived in Wooster, few houses had gas or electricity, and Madison Avenue was unpaved. Without the aid of a telephone or car, Viola welcomed to Wooster the family of each new scientist. Around her, women gathered to sew and exchange news about their children. From this informal arrangement, Viola and her friends created in 1914 a women's group, later named the Viola Thorne Club. Under her leadership the women held picnics, summer barbecues, and Christmas parties. They donated money for school lunches, volunteered in the hospital, and taught Sunday school. Viola Thorne's station—and it belonged to her as well as to Charles—had sunk permanent roots in Wooster and had become as much a family as a workplace.

Viola Thorne, wife of Charles. Like him, she considered the station staff a family. She formed the women's group that later became the Viola Thorne Club.

No community can exist without rules, and Charles Thorne, like Moses, was a lawgiver. He required his men to work ten-hour days, six days a week, and even reserved the right to demand "more than ten hours' work for a few days at a time without compensation." Sunday was no day of rest, for farm hands had to care for livestock and maintain greenhouses, fruit orchards, gardens, barns, and laboratories, and they had to repair implements without pay. Thorne forbade anyone

The station staff in 1895. Front row from left: E.C. Green, W.J. Green, Charles Thorne, J.F. Hickman, and F.M. Webster. Back row from left: Annie B. Ayres and Bertha E. Wildman.

Model Ts and buggies at Wheat Field Day. The man and the toddler in the foreground are unidentified (1918).

The station staff in 1907: forty-six men and women. Charles Thorne is in the middle row, fourth from the left.

from engaging in "unnecessary conversation," cursing, smoking tobacco at work, or drinking alcohol "at any time."

These rules were strict, but then again, a job at the station was no sinecure. A typical farmhand was on the job by 4:00 A.M., though the first few hours did not count because he spent them tending his horse, shoveling coal, oiling a steam thresher, or doing other routine tasks. The real workday began at 7:00 A.M. and continued, with only an hour's lunch, until 7:00 P.M. In return for 15 hours he received $2.50, which beat the daily rate at most jobs.

Although these hours seem draconian today, they were typical of a patriarchal era. Andrew Carnegie gave his men only one day off per year, July 4. Henry Ford inspected workers' homes for cleanliness and polite children. George Pullman required laborers in his sleeping car company to read the Bible every day or, if they could not read, to pray daily.

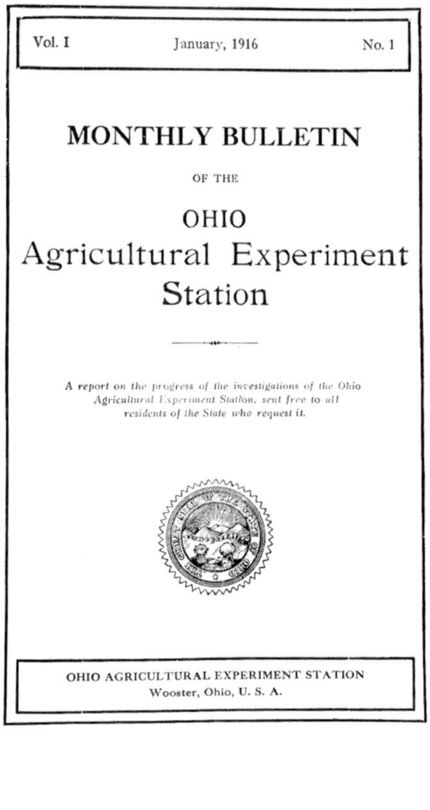

Vol. I — January, 1916 — No. 1

MONTHLY BULLETIN

OF THE

OHIO

Agricultural Experiment Station

A report on the progress of the investigations of the Ohio Agricultural Experiment Station, sent free to all residents of the State who request it.

OHIO AGRICULTURAL EXPERIMENT STATION
Wooster, Ohio, U. S. A.

The *Monthly Bulletin*, started in 1916, carried the station's findings to farmers. Subscriptions were free. Ohio farmers read it. So did people from far away. Among them: Eugene W. Hilgard, director of the California Agricultural Experiment Station.

But Thorne was no slave to the clock. A story has it that one day he entered the town's barber shop for a trim, only to find a greenhouse foreman in one of the chairs. All became silent while Thorne decided what to do. With a casual air, Thorne remarked that because the man's hair had grown on state time it could properly be cut on the clock. Before anyone rushes off to the barber shop, however, be forewarned that Thorne later reversed this decision: get your hair cut on your own time.

If this story is fiction, so much the better. Thorne was less interested in being a taskmaster than in bringing the lessons of agricultural science to every farm in Ohio. Whereas Lazenby had issued only an annual report, Thorne produced an avalanche of bulletins, monthly and bimonthly research summaries, and circulars. He charged

Carlos Grant Williams, an agronomist who joined the station in 1903 and his closest collaborator, with writing a column, "Field Crop Notes," for *Ohio Farmer*. Like Lazenby, Thorne spoke with farmers at their institutes, the Ohio State Fair, the Ohio Agricultural Convention, and meetings of the Horticultural Society. He invited farmers to tour the station and created Wheat Field Day in 1917 for this purpose.

Equally important, Thorne was a pioneer in the movement for agricultural extension, which had roots at Ohio State. There, two students formed the Agricultural Students Union in 1895 with help from Dean Thomas F. Hunt and Augustine D. Selby, a botanist at the station. The four set up practical demonstrations for farmers and used seed Thorne provided. This work needed a formal structure, and in 1904 Thorne hired L.H. Goddard to organize the Department of Cooperative Experiments at the station.

A.B. Graham, the founder of 4-H and who, along with Charles Thorne and L.H. Goddard, pioneered agricultural extension in Ohio. Its success led Congress to fund extension nationwide through the Smith-Lever Act of 1914.

A year later Hunt appointed A.B. Graham, a former school superintendent, as the first head of extension. In addition to creating the forerunner of the 4-H clubs, Graham joined with Goddard and Thorne in creating a farm demonstration program. William Oxley Thompson, president of Ohio State, used their program as a model in proposing a nationwide extension service. This idea formed the core of the Smith-Lever Act, which Congress passed in 1914 and which established the Cooperative Extension Service.

An Ohio 4-H club band. The photograph dates from the 1920s.

A station wheat test plot. Well into the 20th century, wheat was grown on more acres than any other crop in Ohio except corn.

William J. Green, standing, addresses visitors in 1897. Green was the station's first horticulturist and vice director.

William J. Green slices an apple in search of mouldy core, a fungal disease (1898).

The station mail room in 1916, filled with incoming letters from farmers and scientists and outgoing bulletins from Charles Thorne and staff.

Charles Thorne in a corn field (1897). His research on crop rotations and fertilizer and manure use brought the station a national reputation.

Applying lime to a test plot. Charles Thorne followed the example of the Rhode Island Agricultural Experimental Station in applying lime to acidic soils. The Bank Barn is in the background (circa 1895).

J. Freemont Hickman, the second superintendent of field operations. The rigid pose and serious demeanor are typical of 19th-century portraits.

The extension service gave Thorne a vehicle for unveiling the latest science. Dearest to his heart was the work in restoring soil fertility. During the early twentieth century he created the nation's most extensive experiments with crop rotations and fertilizers. To be sure that these experiments would have value for farmers throughout Ohio, he established fourteen substations, with at least one on each major soil type. These substations, he boasted, made all of Ohio his laboratory.

This work brought Thorne national acclaim. In 1904 Cyril Hopkins, the celebrated agronomist at the University of Illinois, remarked that "I have gotten more information of value to Illinois from the study of soil work that has been done in Ohio, than I have from that of any other state." In Washington, D.C., in 1909, Milton Whitney, head of the U.S. Department of Agriculture's soils bureau, praised Thorne's work as the most thorough in the United States. At the annual meeting of the American Association of Agricultural Colleges and Experiment Stations in 1916, Frederick B. Mumford, dean of the agricultural college at the University of Missouri, rose from his seat to nominate Thorne to become the group's president. Mumford remarked that Thorne's work stood as a monument to agricultural science and that Mumford could add nothing to it with a nominating speech. This being so, he resumed his seat without further comment, the gathering of scientists and college administrators unanimously agreed, and the next year Thorne served as president. *Ohio Farmer* called him "the grand old man of agriculture" and remarked that farmers everywhere applied his lessons in their fields. Farmers and extension agents from all over America wrote to Thorne for advice. By his own admission, he had taught "the acre to produce

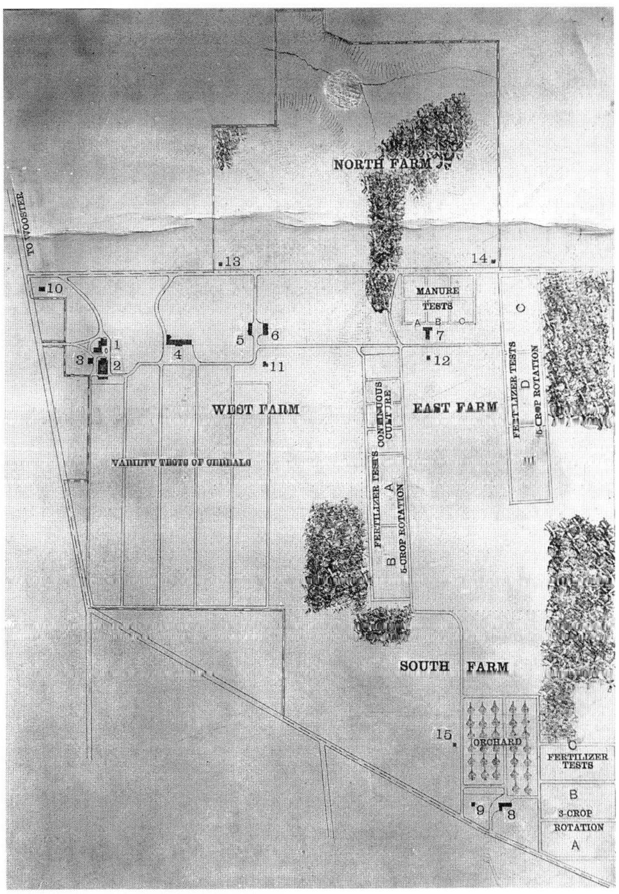

A map of the station in 1897. The road marked "To Wooster" is today's Madison Avenue.

more food." He had, farmers and scientists knew, become the Thomas Edison of the farm. His discoveries illuminated the principles of restoring soil fertility. In the process, Thorne helped transform agriculture from an ad-hoc enterprise into a science, a triumph that today fills every grocery shelf.

But abundant crops meant a feast for insects. To combat them, entomologists at the station sprayed solutions of lead arsenate and other toxins on grain fields and fruit trees with the idea that a hungry insect would poison itself in attempting to get a free meal. William Chamberlain allowed entomologists to test their sprays on his apple orchard and took every opportunity to praise his apples, free of insect and worm damage, as the best since Johnny Appleseed's day. By 1917 orchardmen were reporting as much as $400 profit per acre of apples, thanks to these insecticides. On a grander scale, in 1921 entomologist Joseph S. Houser supervised the first insecticide application by airplane. Afterward he boasted of millions of dead caterpillars, *National Geographic* published a story on this new use of aviation, and crop dusting immediately became a popular practice.

No less troubling than insects were pathogens. Fortunately for Ohio farmers, Thorne hired Augustine Dawson Selby in 1894 as botanist and chemist. Before the decade's end, Selby chaired both departments and had made the study of crop diseases his primary task.

The first aerial crop-dusting experiment, conducted by station scientists and U.S. Army pilots over a catalpa grove in Troy, Ohio, August 3, 1921. The test used powdered lead arsenate. It killed more than 99 percent of its target: catalpa sphinx caterpillars that were eating the leaves of the trees. After the first pass the pilot saw the scientists "fleeing for their very lives out the far side of the grove to escape the cloud of poison." The photograph was taken from a circling DeHaviland airplane.

At right is 1st Lt. J.A. Macready of the U.S. Army, who piloted the plane during the first aerial crop-dusting experiment. The man at the left is unidentified but is possibly a "Mr. Dormoy," who operated the crank on the dusting hopper. The biplane is a Curtiss J-N6 "Jenny."

In contrast to Thorne, Selby sought close ties with Ohio State by strengthening the land-grant mission there. As an undergraduate, Selby helped form the Agricultural Students Union in 1895. Once in Wooster, he was the first to send scientists to the university for additional study and graduate degrees. Because he viewed the university as a training ground for his scientists, he did not hesitate to ask the legislature to fund it. Indeed, his efforts helped finance construction of the university's Horticulture and Forestry Building. At the same time, he joined Thorne in prying money from state coffers for the station's branch farms.

As a scientist, Selby was a pioneer in Ohio. He was the first at the station to identify himself as a plant pathologist on his letterhead—a scientist who studies plant diseases and plant disease management. In doing so, he was nearly seventy years ahead of his time, for only in 1967 would the station formalize the Department of Plant Pathology. Furthermore, he was among the first in Ohio to embrace Mendelian genetics for describing the mechanism of heredity. His weed manuals and his *A Condensed Handbook of Diseases of Cultivated Plants in Ohio* were encyclopedic in their coverage and served as college textbooks even after his death.

Selby was equally adept in organizing and promoting scientific societies. He was a charter member of the Ohio Academy of Science and a frequent speaker at its meetings. At the national level, he helped found the American Phytopathological Society and its journal, *Phytopathology*. In these labors, Selby, like Thorne, built a national reputation for himself and for the station.

An early pesticide spray rig being tested on university land in Columbus (circa 1900).

At right, William Oxley Thompson, president of Ohio State, tours the station's chemistry laboratory (1899).

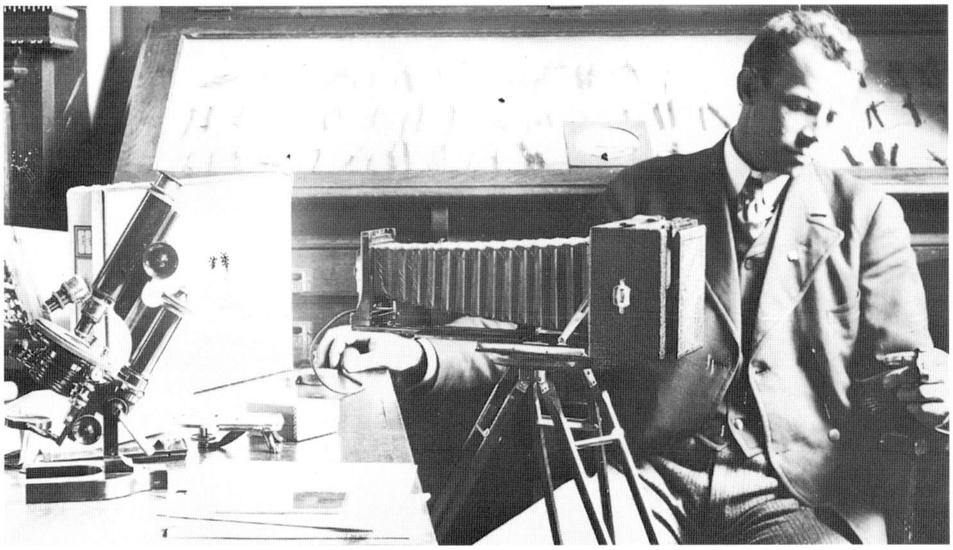

W.H. Goodwin, an entomologist, mounts and photographs insect specimens (1910). Charles Thorne thought photography was vital for science and promotion. He set aside the second floor of the chemistry wing of the Administration Building as a photo lab.

He might have done even more had he been located at a university and had respiratory problems not forced his retirement in 1923. Today, the station's plant pathology building, Selby Hall, remains a monument to his achievements.

Of course, crop growers were not alone in reaping rewards from the station. In 1907 Ernest B. Forbes joined the station as chief of the new Department of Nutrition. His work strengthened the sensible belief that hogs needed a calcium-rich diet to maintain bone density. A grain-based diet was not enough, and so Forbes promoted the value of adding soybeans and alfalfa to hog rations. His experiments confirmed, moreover, that dairy cattle needed mineral-rich rations because they lost minerals from their skeletons during milk production. After scientists at the experiment stations in Wisconsin and Connecticut discovered the first vitamin in 1913, Forbes began supplementing livestock rations with vitamins. His work reinforced the value of feeding livestock a diet rich in protein, carbohydrates, fat, minerals, and vitamins. Forbes' legacy to the station, unfortunately, rested less on research than on a clash with an animal scientist that revealed the limit of Thorne's authority.

5 THE LIMIT OF AUTHORITY

THAT SCIENTIST WAS BERTON E. CARMICHAEL, WHO GRADUATED FROM THE University of Illinois in 1905 with a bachelor's degree in animal husbandry. Thorne hired him that year as chief of the new Animal Husbandry Department in the hope that he would energize animal science at the station, which was in the doldrums.

The lethargy of livestock research was partly the result of bad luck. In 1899, bovine tuberculosis swept through the station's dairy herd, killing seventeen cattle. Thorne lamented the catastrophe as an incalculable loss and admitted that he did not expect the legislature to grant him funds to replenish the herd. To make matters worse, fire consumed the dairy laboratory and creamery in 1913; fortunately, the cows were in their barn, a separate structure, and suffered no losses. Thorne did not rebuild until 1915. Perhaps he was unable to obtain sufficient appropriations until then, or perhaps he had not perceived an immediate need to erect a new lab and creamery. Either way, these setbacks did nothing to invigorate animal research at the station.

While livestock research languished, Thorne led a crusade against soil exhaustion that won the station national acclaim. Like the Roman farmers of antiquity, Thorne could plant his feet into Ohio's soil and know that its fertility was his doing. Carlos Williams, who joined the station in 1903 as an agronomist, was hard at work breeding new wheat varieties for farmers. Entomologists had declared war on insects and slugs and would soon soar through the air in their quest to rid wheat, corn, and oat fields of insects.

In contrast to these gains, animal science seemed feeble. Stockmen had not been able to bask in Thorne's scientific triumphs. In this difficult time, Thorne expected grand deeds from Carmichael. The year Thorne appointed him was a critical time, for only then did Thorne publicly admit that stockmen had been criticizing the station for ignoring their interests. He attempted to defuse the attack, first, by blaming the legislature, which had never given the station enough money to pursue livestock research. This assertion may have had merit. Since Townshend's failure in 1880 to cajole the legislature into creating an experiment station with a $10,000 appropriation, no one had pried much money from it for agricultural research. Second, Thorne argued that the station could best serve farmers by tackling their most intractable problems. Everyone knew, therefore, that he was obliged to pour his energies into helping farmers improve soil fertility.

Berton E. Carmichael, chief of animal husbandry, 1907–1921. His feud with Ernest B. Forbes, chief of nutrition, frustrated Charles Thorne and plunged the station into a bitter lawsuit.

If Thorne could not spare himself in the service of stockmen's problems, then this responsibility fell to the new animal husbandry chief. Upon arriving in Wooster, Carmichael initially worked "with great enthusiasm," recognizing his opportunity to create an animal research program where none existed.

Thorne was impressed by Carmichael's diligence and now saw his opportunity to establish, at last, a legitimate livestock research program. In 1907, he hired Ernest Forbes, the station's first full-time scientist with a doctorate degree, to create a Nutrition Department. Forbes appeared to be the right man for the job. He had been a professor of animal nutrition at the University of Missouri, where, in hog-raising territory, he had established a reputation as an expert in the mineral nutrition of swine. Thorne anticipated that Forbes would make the Ohio station a center for hog nutritional studies and, in so doing, win allegiance from western Ohio's hog raisers.

Thorne now believed he had assembled the elements of successful livestock research. He had Carmichael working on the problems of livestock raising in general while Forbes grappled with the mystery of nutrition in particular. At the same time, Thorne, A.B. Graham, and L.H. Goddard were building an extension program that would foreshadow the Smith-Lever Act of 1914 and would enable them to bring the lessons of livestock research to every barn and pasture in Ohio. The prospect of meatier hogs and beef cattle and an endless supply of milk seemed within the grasp of Thorne and all stockmen.

But as everyone knows, well-laid plans sometimes go awry. The root of the station's troubles was Carmichael, who did not share Thorne's enthusiasm for

his new colleague. Carmichael viewed all livestock studies as his fiefdom and Forbes as a rival lord and thus deeply resented that he had to cede most of his livestock nutrition research to Forbes.

Carmichael managed, however, to retain the feeding trials of horses and cattle by persuading Thorne that he was nearing the climax of these investigations. In 1908 he published two bulletins, which outlined the preliminary results of these trials and promised a definitive publication soon.

The ruins of the station creamery after the 1913 fire.

This was an empty assurance, as he thereafter ceased to work productively. Instead he bought sixteen horses, fed them to maturity, and sold them. He next purchased twenty Belgian fillies in order, he claimed, to begin breeding experiments. Although he amassed a large herd during the next decade, he failed to publish even one account of this work.

While presumably engaged in this study, Carmichael asked Forbes, in late 1916 or early 1917, for permission to use part of his laboratory to investigate "the chemical composition of meat." This was a clumsy incursion into Forbes' domain, and Forbes refused, citing lack of laboratory space and remarking that even with space he would have objected because Carmichael "had not shown the courtesy and consideration essential to harmony in the joint use of the same equipment."

Forbes then complained to Thorne, who intervened. On January 18, 1917, Thorne warned Carmichael to drop his attempt to encroach on Forbes' turf, reminded him of his promise in 1908 to conclude his feeding trials with horses and cattle, and asked Carmichael to return to his research. This plea, couched in the most tactful language Thorne could muster, failed to impress Carmichael. Instead he fired off a letter to Forbes the next day in which, making no reference to Thorne's admonitions, he accused him of having used improper influence in determining the location of his laboratory.

Forbes again approached Thorne and now asked him to silence Carmichael, a request that put Thorne in a difficult position. Years later Thorne conceded that he should have then fired Carmichael because of his refusal to work and for his attack on Forbes. But Thorne was so eager to build an animal science program that he held his tongue. Thorne also did not fire Carmichael because he feared the public might protest the dismissal of a departmental chief.

Thorne's attempt to weather the storm appeared to have succeeded when, in January 1918, Forbes accepted a commission to serve in World War I as a major in the U.S. Army sanitary corps. Thorne believed that with Forbes an ocean away for the foreseeable future, the feud would subside. Thorne even hoped that Carmichael would finally return to work and that he would be able, at last, to boast a legitimate livestock research program.

Instead, in a letter to the sanitary corps that February, Carmichael accused Forbes of punching a subordinate. Worse, in the midst of America's crusade against the Kaiser, Carmichael disparaged Forbes as an unpatriotic lout who had refused to buy Liberty Bonds and had never attended the parades for Wooster's soldiers.

Although these charges did not dissuade the Army from commissioning Forbes, they made public the squabble between Carmichael and Forbes. Thorne was outraged and summoned Carmichael to an emergency session of the Board of Control on March 1, 1918. There, Carmichael impressed no one by hiding behind the excuse that he had hoped to protect the Army from unknowingly commissioning an unpatriotic officer. The next day Thorne suspended Carmichael without pay, and on March 7, the board fired him.

Carmichael immediately appealed to the Ohio Civil Service Commission for reinstatement. At a hearing on May 21, he argued that the board had unfairly discharged him for making factual statements to the Army. Carmichael was at his best, or worst, depending on one's viewpoint, at this hearing, displaying a virtuoso ability to clutter an issue with red herrings and to impugn others on the pretext of

Station workers plant corn in a test plot. The year the photograph was taken is unknown but was sometime during the 1920s or 1930s.

faithfully following orders. He now asserted that he had only mailed his letter attacking Forbes after Thorne had reassured him that it was his duty to do so. This statement was surely a lie, and when Thorne, who represented the Board of Control at this hearing, objected, Carmichael shifted the ground to Forbes' character. The fact that his character was irrelevant to the hearing did not deter Carmichael, who now enlarged his original assertions by claiming that Forbes had taken a small plot at the station for his own use on the pretext of making a Victory Garden. Worse, Forbes had pleaded poverty when refusing to buy Liberty Bonds though he had sufficient resources to drive a Buick. This was a deft stab; Woosterites were plain folks who drove Fords and viewed Buicks as extravagant. Finally, Carmichael asserted that Forbes was so greedy that he had attempted to draw his full salary from the station while serving in the Army.

For his part, Forbes had made himself an easy target for these charges. He had indeed struck a subordinate around 1910 over a minor incident but had immediately apologized to him. The charge of double-dipping, though a bit more complicated, was essentially true. Forbes had asked Thorne to continue paying him some portion of his salary so that his family would have adequate income during his absence. Although this would have been legal, Thorne had refused on the grounds that it was neither ethical nor in the station's financial interest. Equally serious, Forbes had clumsily attempted to pose as an authority on the nutrition of human infants. He had joined the vitamin craze that followed the discovery of the first vitamin in 1913. In 1916, he speculated that cows' milk plus the proper combination of vitamins constituted the perfect nourishment for human infants. But Forbes was now blundering into an area where he had no expertise. Worse, he had apparently failed to appreciate the obvious implication of this idea. If vitamin-fortified cows' milk was nutritionally superior to mothers' milk, then women had no business breast-feeding their babies. Area farmers recognized that this implication undermined the traditional female role as nurturer and were indignant; they wondered aloud why

A parking area full of cars at the 1918 Wheat Field Day. In the background is the Bank Barn, in the foreground, corn.

Students and a teacher check hogs at an unidentified location in the 1930s.

their tax dollars should support animal research that was so clearly subversive. Unwittingly perhaps, Forbes had weakened support for livestock research and discredited himself at a time when the animal sciences were new and thus vulnerable at the station.

At the end of a day of testimony, Carmichael had succeeded in casting Forbes, rather than himself, as the villain, and the commission promptly reinstated him, deeming his attack on Forbes as entirely appropriate. A crestfallen Thorne complained to Governor James Cox that "the outcome of this decision is that a man who has discredited himself ... is permanently fastened upon the payroll of the state."

Since Thorne could not fire Carmichael, the thought occurred to him that he might abolish his position. On January 15, 1919, he informed Carmichael that the board was consolidating Animal Husbandry and Nutrition on February 1 and that Forbes would assume charge of the new Animal Industry Department. Thorne justified this move on January 19, announcing in *The Wooster Republican* that the board was obliged to unify these departments since work in Animal Husbandry had been "in abeyance for ten years."

But Thorne had miscalculated. Carmichael, persistent as ever, sought a restraining order from Lawrence Critchfield, judge of the Wayne County Common Pleas Court, to thwart the merger. His argument that the board members had conspired to remove him persuaded Critchfield, who on January 31 temporarily enjoined the board from uniting the two departments.

The ruling provided Carmichael with only a brief respite, since Thorne ordered him on May 31 to leave his residence in one of the station's houses on the pretext that "it was urgently needed for the housing of men whose duties made it important that they should live near their work." Carmichael was vulnerable to this salvo; since his arrival in Wooster he had lived on the station's grounds, an economy prompted by his lack of an automobile and the dearth of inexpensive housing in town. Desperate, he filed suit against Thorne and the other board members in the same court on June 6, asking Critchfield to permanently prohibit them from colluding to deprive him of work and lodging.

The hearing, which began on July 22, 1919, and continued until December 30, 1920, was an unedifying spectacle. On October 2, 1919, Critchfield held Thorne in contempt and fined him $100 for refusing to answer whether Carmichael was still the chief of Animal Husbandry. For his part, Carmichael could not explain why, since his reinstatement in May 1919, he had proposed fifteen different lines of investigation, had never concluded the feeding trials of 1908, and had not published a single bulletin since then.

Instead, he again relied on red-herring and ad-hominem attacks to rescue him from his ineptitude. On November 13, 1920, at Carmichael's behest, Dwight Lincoln and Grant Case, secretary of the American Rambouillet Sheep Breeders Association and president of the Ohio Poultry Association, respectively, filed a joint lawsuit against Thorne, charging him, first, with incompetence stemming from "an uncontrollable mania to be revenged upon any employe[e] or person who differs from [*sic*] him." Second, and more serious, they alleged that Thorne had misused public funds. At this point the breeders association produced the audit of fiscal year 1918, in which John Bliss, the chief examiner, uncovered overpayment of one worker, increased pay of another to cover medical expenses, and a litany of alleged waste, abuse, and theft, including the disappearance of a watermelon. Third, they accused Thorne of hindering the progress of animal research at the station. This work had been "in a chaotic condition," they asserted, until Carmichael had determined to build a first-rate program; but Thorne, it was claimed, had harried him at every turn.

These allegations were silly. With the exception of Carmichael, those who knew Thorne (and Lincoln and Case probably did not), from farmhands to administrators at the U.S. Department of Agriculture, lauded his integrity.

The auditor of state investigated Bliss's report and concluded that Thorne had not misused any funds. Indeed, as the example of the stolen watermelon underscores, Bliss was an overzealous auditor who had only magnified the inconsequential. Finally, Thorne had attempted to coax Carmichael as tactfully as he could to return to work. Carmichael, rather than Thorne, had undermined animal research.

The men in the front row guided the station during its first fifty years. From left, Carlos Grant Williams, who succeeded Charles Thorne as director (1921–1937); Augustine Selby; William J. Green; and Charles Thorne.

Critchfield saw the absurdity of these charges and dismissed the case on December 30. But on that day he also ruled in Carmichael's favor, permanently prohibiting Thorne and the other board members from unifying Animal Husbandry and Nutrition and from removing Carmichael from his residence. The board had undertaken such action, Critchfield ruled, to rid the station of Carmichael rather than to unify two departments. But the Ohio Civil Service Commission had earlier reinstated Carmichael, and he could not now be fired without cause.

At this juncture Thorne passed the directorship over to Carlos Grant Williams, the chief agronomist since 1906 and a close friend and collaborator in Thorne's experiments on improving soil fertility. The tumult over Carmichael had delayed rather than hastened this action. In February 1918, Thorne had asked to retire at the end of the year, conceding that the administrative burdens of being director required a young man's vitality, whereas he was nearly seventy-two. Retirement then would afford him time, he hoped, to summarize his lifetime of experiments with crop rotations, fertilizers, and manure in a single publication. Thorne suggested that the board appoint Williams as his successor and allow Thorne to resume charge of the Soils Department. In view of Carmichael's antics, however, the other board members asked Thorne in March 1918 to shepherd the station through this crisis. Thorne agreed only to a temporary extension of his tenure, and in January 1920 Thorne renewed his request to retire, now at the end of 1920. This time the other board members acquiesced, and on December 30, 1920, Thorne handed the reins to Williams.

Thorne's retirement did not mean that Carmichael had won. Now chief of the Soils Department, Thorne could not endure the spectacle of his nemesis permanently thumbing his nose at him. Thorne immediately filed an appeal, and the Ohio Court of Appeals finally saw Carmichael's behavior for what it was: insubordination and dereliction of duty. On March 21, 1921, the Court of Appeals upheld the board's decision to consolidate the two departments and ordered Carmichael to leave his residence or face trespassing charges. The next day a chastened Carmichael returned his keys to Williams and plunged into obscurity. At last, after haggling with Carmichael for more than a decade, Thorne prevailed.

But victory came at a high cost. The Carmichael case had shown, as everyone knew, that not even Thorne could control everything that happened at the station. His efforts to strengthen livestock research had taken him away from his crusade against soil exhaustion and the chinch bug and had darkened the twilight of his career. Yet these efforts had yielded nothing. The animal scientist he welcomed to Wooster in 1905 proved to be dishonest, disloyal, and inept. To put it plainly, Carmichael was a virtuoso rogue and a poor scientist. In Carmichael's sixteen-year career he disgraced himself and brought unflattering press to the station. Forbes never recouped his reputation and resigned at the end of 1920. In the wake of Carmichael's antics, animal research lost

Charles Thorne, back row, fifth from the left, with a group of Wooster bankers about 1921. In addition to being a scientist and administrator, Thorne was president of the Wayne Building and Loan Company. *(Photo courtesy of Wayne Savings Community Bank.)*

rather than gained ground on the plant sciences. Not until Leo L. Rummell, a journalist, businessman, and agricultural enthusiast, became director in 1948 would the station have a chief who would admit an interest in livestock research. In the short term, the Carmichael case made clear that Thorne's long tenure had come to an end. The station's future, though it would always bear the stamp of Thorne's influence, lay with others.

6 TO BE OR NOT TO BE

Workers run a threshing machine about 1925. Note the absence of rubber tires. Station scientists conducted the first tests of pneumatic tractor tires in 1932; steel-lugged wheels were soon replaced. The Department of Agronomy used the building in the background until it was torn down around 1980.

DIRECTORS SINCE THORNE HAVE FACED THE DILEMMA OF HOW TO RETAIN THE strengths of his era while changing the station enough so it could respond imaginatively to the needs of a new generation. No one knew this dilemma better than his immediate successor, Carlos Williams. He followed Thorne to the director's office because he was the closest thing to a Thorne clone. Williams was, after all, a careful agronomist and a clear writer, as Thorne had been. And Williams made plain his connection to Thorne by dedicating the entomology building to him in 1927. Like Thorne, he corresponded with farmers and chatted with them at Wheat Field Day and the Ohio State Fair. He spoke of their noble labor and his pride in serving them through science. Williams was as good a successor as Thorne could have wanted.

But as everyone knew, Williams was no

carbon copy of Thorne. Whereas Thorne could rumble like a piece of heavy artillery, Williams preferred to communicate with a quiet word or friendly nod. Thorne accepted conflict as an inevitable part of being director, but Williams favored negotiation and compromise. On the one hand, Thorne was a realist who met everything head-on. His prose was equally direct and at times even terse. He had no use for flowery effect and probably never read a poem on purpose. On the other hand, beneath his Brooks Brothers suit and behind his spectacles, Williams had a poet's soul. His poems fill a small volume, copies of which reside in the center's archives and in the Wayne County Public Library. Williams' poetry described nature's splendor and his delight in the ordinary events of daily life. In "On Seeing Some Workmen Building Some Steps," he wrote:

I saw four men at work on a wee job of brick laying. Five steps there were.
(I wonder why not a man for each step?)
First, a little earth had to be dug away and placed at the side.
One man dug this and three looked on.
I think they wished him well, but I do not know. Then came some measuring and leveling.
One man did this, too, while the others looked on.

In his own quiet way, this poet knew how to connect with farmers. Williams had grown up on a dairy farm in Trumbull County, Ohio, and had spoken at Farmers' Institutes before he joined the station. He described farming as a business and farmers as corporate executives. Such language flattered farmers, who felt a kinship to Williams throughout his career.

The poet in Williams no doubt found pleasure in the Firestone Gardens, which no longer exist; they were located between Thorne Hall and Madison Avenue. In 1925, the Board of Control purchased these gardens, then in their prime of color and fragrance, from Frank and Mabel Firestone. The gardens encircled more than an acre, on which human effort had tamed nature,

A tractor powers a threshing machine. Just as the Model T supplanted the horse-drawn buggy, the tractor replaced the horse on many farms in the 1920s.

An aerial view of the station about 1925. In the center are the Firestone Gardens, which Frank and Mabel Firestone sold to the station in that year. The station agreed to maintain the gardens for five years; it did so for more than thirty.

bringing forth turf grass, red and yellow tulips, red and yellow roses, orange marigolds, daisies, and hydrangeas. At the center of the garden, a pond and bird bath attracted geese and ducks to feast on bread crumbs strewn by the locals. The purchase agreement required the station to maintain the gardens for five years, though its staff kept them in bloom more than thirty years.

Williams spoke in verse of workmen who "looked on." Perhaps in some metaphorical sense, Thorne continued to look on the station after his retirement. Even so, it could not remain unchanged, in part because Williams was not Thorne.

A second reason the station could not avoid changing was because Williams favored a closer relationship with Ohio State than Thorne thought was wise. In 1925, Williams and the other members of the Board of Control agreed with the trustees to exchange faculty as needed. Under the terms of this agreement, university scientists could do research at the station and receive part of their salary from it. At the same time, those in Wooster could draw pay from the university by teaching courses in Columbus. That year, the departments of Agricultural Engineering and Home Economics at the university and at the station exchanged faculty and thereby laid a foundation for the unification of the agricultural college and station that took place nearly sixty years later.

The third change was that Williams' station was no longer simply a man's world. Whereas in 1900 only two women were on staff, by 1925, fourteen women worked at the station. They wanted an organization, but the Viola Thorne Club offered membership only to women whose husbands worked at the station. So the station's working women did what Viola would have done—they created their own organization, the Girls' Club, in 1926. It began as a forum through which women discussed their work. But like the Viola Thorne Club, the Girls' Club grew into a social agency with ties throughout

Wooster. Its members volunteered for the Red Cross and the Human Services Center and helped fund the Christian Children's Fund and Give-A-Christmas.

Fourth, Williams could do little more to raise yields by enriching the land because Thorne had done so well in restoring soil fertility. Instead, Williams concentrated on breeding higher-yielding crops. In doing so, he changed the station's emphasis from the soil to the plant.

From left: Glen H. Stringfield, R.D. Lewis, and D.F. Beard in 1940. All were agronomists. Stringfield developed inbred and hybrid corn that resisted European corn borers, saving Ohio farmers millions of dollars.

Williams was the right person to lead the station in this direction, for as an agronomist he had bred three widely grown wheat varieties: Trumbull, Gladden, and Portage. He needed someone to do the same for corn, and in 1924 the U.S. Department of Agriculture filled that need by assigning Glen Herbert Stringfield to the station.

That year the lanky Nebraskan had earned a bachelor's degree in agronomy from the University of Nebraska. In both aptitude and training, Stringfield was the quintessential corn breeder. He exalted corn as the "king of American crops" and believed that his work strengthened "our faith in agricultural research." More than that, Stringfield believed he was helping to expand "the frontiers of knowledge." He saw his work as a quest for truth, which he pursued with an intensity that was akin to Thorne's passion.

During thirty-five years at the station, Stringfield derived nearly 200 inbreds and hybrids resistant to the European corn borer. Thanks in part to Stringfield, yields per acre of corn more than doubled during his years at the station. Between 1954 and 1965, borer damage to Ohio corn diminished from $8,500,000 to $600,000. Indeed, one participant in the plant breeding symposium at Iowa State University in 1966 remarked, "The success of the breeding program [to derive borer-resistant hybrids] is indicated by the fact that in Ohio the borer is no longer a major problem."

Edmund Secrest, who succeeded Carlos Williams as director (1937–1947). Before coming to the station he was a protege of Gifford Pinchot in the U.S. Department of Agriculture Forest Service.

Edmund Secrest established the Forest Arboretum in 1906. Today it is named in his honor. It attracts an estimated 10,000 visitors a year.

That year Governor James A. Rhodes recognized Stringfield's service to farmers by naming him as one of the first three men to be inducted into the Ohio Agricultural Hall of Fame. Stringfield was the only one of the three who was not a native Ohioan, suggesting that Ohio farmers had adopted him as one of their own. "Likely no other man," a colleague remarked about Stringfield after his death, "was responsible to such high degree in increasing food production."

Long before Stringfield finished his work, Charles Thorne died on February 29, 1936, just a few months short of his 90th birthday. At his birth, the Republican party did not exist, and few people had heard of Abraham Lincoln. Not a single railroad spanned the country. Most people made their rounds on foot or by horse. In science, virtually no one in the United States besides Asa Gray at Harvard University knew of Charles Darwin, and the world's only agricultural

Joseph S. Houser, the entomologist who led the crop-dusting experiments. The microscope and the mounted insect collections were typical of any entomologist's office (circa 1930).

experiment station was in Rothamstead, England. No Ohioan had yet thought of creating one.

At Thorne's death, Franklin D. Roosevelt was in the midst of the second New Deal. Under Roosevelt, the Social Security Administration dispensed money to retirees, the Federal Deposit Insurance Corporation protected bank deposits, and the Rural Electrification Administration was stringing electric lines through the countryside. By then, Americans dashed across the land by car and soared through the air by plane. Both horse and railroad had receded into the memory of most Americans. High-school biology textbooks in Europe and in the United States taught that humans and apes shared a common ancestor and had thereby angered creationists in Dayton, Tennessee. Agricultural experiment stations had spread throughout Europe and the United States, and in Ohio, the station received an appropriation of just under $500,000 in the year Thorne died.

Edmund Secrest, whose fiscal caution as director led one scientist to call him a "penny pincher."

Despite his age, no one had expected Thorne to die because he was still so vigorous. When Thorne had retired fifteen years earlier, he had vowed to summarize his lifetime of experiments in restoring soil fertility. At age eighty-six, he fulfilled this promise, publishing his magnum opus, *The Maintenance of Soil Fertility*. This work closed a publishing career that Thorne had begun more than sixty years earlier as a free-lance writer. Along with Michelangelo and Giuseppe Verdi, Charles Thorne lived a life of creativity even as an octogenarian.

At his funeral, the officiating minister claimed Thorne's soul for the Presbyterian faith, but it also belonged to farmers and scientists. Condolences poured into the station from farms throughout Ohio, from the U.S. Department of Agriculture, and from experiment stations and land-grant colleges from throughout the country. Throughout its history, the station has benefited from innumerable brilliant people, but it has had only one Charles Thorne, and it had him for nearly half a century. Today the station lives with his memory and his challenge to continue serving farmers, even as it expands its mission to include urban and suburban Americans.

A year after Thorne's death, Carlos Williams handed the directorship over to Edmund Secrest, a Kansan who was a product of a conservation movement that had deep roots in America. As early as 1872, Congress had set aside Yellowstone as the first national park, and in 1881, created the Division of

Staff and guests display the first fistulated steer, developed in 1937 to study bovine digestion and to improve nutrition. The plug allows access to the rumen and its contents. The technique is still in use today, and visitors often ask to see "the cow with the window in its side."

Fords, Chevrolets, Plymouths, Buicks, and Nashes at a station field day in 1940 or 1941.

Forestry within the U.S. Department of Agriculture. When Theodore Roosevelt became president in 1901, he brought to the White House a hatred for "a man who would skin the land." In a flurry of activity, Theodore Roosevelt created fifty federal wildlife sanctuaries and five national parks, persuaded Congress to name the Grand Canyon a national monument, and barred loggers from 172 million acres of public land. Thanks to Roosevelt, conservation came into its own. And, as his lieutenant in this movement, Roosevelt had, in 1901, appointed Gifford Pinchot to head the Division of Forestry, which by then had been reclassified as a bureau. In 1906, Pinchot sent Secrest to Wooster to establish an 85-acre forest preserve under the joint management of the bureau and the station.

There, Secrest, as Williams had done, formed a friendship with Thorne, though the chief forester's marriage to an heiress of the Hoover fortune propelled much of his rise to prominence. His scientific accolades followed in due course. Twice Secrest received an honorary doctorate, once from his alma mater, Kansas State College, and again from The College of Wooster. Between 1925 and 1937, Secrest presided over the Ohio Forestry Association. At other times, he was a member of the Wayne County Board of Park Commissioners and the Ohio Conservation Council, as well as a trustee of Dawes Arboretum and The College of Wooster.

Being well-connected, Secrest could afford to be less concerned with political realities to the point that he tried to make forestry the equal of agronomy at the station. This agenda countered the desire of farmers to clear land for farming and grazing and made Secrest the first director to enlarge the station's mission beyond agriculture. Indeed, he appealed to an urban middle class that wanted Ohio to retain the beauty of its woodlands.

Even when the station concentrated on agriculture, it did more than help farmers solve production problems. At the end of World War I, farm prices fell by more than two-thirds. Farmers in record numbers defaulted on their loans, and rural America tottered near collapse. In the hope that the experiment stations could help farmers improve their lives, Congress passed the Purnell Act in 1925. It gave each station $20,000 in additional federal funds to establish departments of home economics and rural sociology. More broadly, the act supported research that held the promise of improving rural living conditions. In Ohio, the Purnell Act funded research by William Krauss, a dairy scientist who joined the station in 1926. During the 1930s, he led a team of scientists at the station and at the Case Institute of Technology in proving that vitamin A deficiency in humans degenerated the optic nerve. He also demonstrated that vitamin D was essential in maintaining bone density. This work led to the vitamin-fortified milk that all Americans consume.

A 1930s threshing machine with rubber tires and belts.

Thanks to Krauss, families no longer fear or even know of rickets.

Under Secrest, scientists not only began to do a new type of science, one with an explicit social content, they began to define research in a new way. Although the current motto that "research begins not in the laboratory, but in the library," is too strong to describe science under Secrest, the expansion of journals during the first quarter of the twentieth century required scientists to spend more time hunting through periodical stacks than they ever had in the past. A library had been at the station since 1888, but it was part of the director's office, and the librarian was little more than the director's clerk. This arrangement would no longer do, and in 1928, Secrest gave the library two rooms on the second floor of the Administration Building. For the first time, the library held an office, reading room, and stacks. To be sure, departments would continue to maintain their own collections until 1968, but the library had taken a step toward

Station workers test a silo filler by feeding corn into it. The photograph was taken in the 1940s.

becoming the station's repository of scientific knowledge. Like universities throughout the United States, the station was becoming a modern research facility.

Of course, the station still focused on traditional areas of research, as Leonard J. Alexander would have been quick to say. He came to Wooster in 1930 with a bachelor's degree in agricultural economics and botany from Louisiana State University. Four years later he received a doctorate degree in plant pathology from the University of Wisconsin.

At the beginning of his career, fusarium wilt and tomato mosaic virus ravaged tomatoes grown in greenhouses. Tomato growers came to Alexander wringing their hands in despair at the damage from these diseases, and in Alexander they found the person who could solve their problems. During four decades at the station, he discovered genes resistant to both diseases and bred some fifty hothouse varieties that saved growers millions of dollars. As late as 1992, greenhouse tomatoes contained the disease-resistance gene that Alexander discovered.

Animal scientists were busy as well. During the Williams and Secrest years, nutritionists compared rations of oats and corn for horses and developed a grass silage that increased milk production in

Leonard Alexander in a station greenhouse. In addition to his pioneering research on disease-resistant tomatoes, he was one of the discoverers of maize dwarf mosaic virus. A plant pathologist, he worked at the station from 1930 to 1970.

Leonard Alexander, right, records the weight of tomatoes about 1966. A meticulous recordkeeper, he developed tomato varieties that resisted fusarium wilt, tomato mosaic virus, or both—a breakthrough for Ohio greenhouse growers.

dairy cattle. They fed chickens rations with varying proportions of alfalfa and corn, and animal breeders crossbred French Charolais bulls with Hereford and Angus females to produce beef cattle with large cuts of high-quality meat.

Williams and Secrest gave the station much to celebrate. They proved that it could do well even without Thorne's charisma. Upon his retirement in 1947, one year after Williams' death, Secrest could be proud that the station's future never looked brighter.

But at the same time the station faced new challenges. The United States had emerged from World War II as the world's most powerful industrial economy. Many Americans were at least a generation removed from the farm and knew nothing about the troubles that farmers faced. These Americans cared only that food was available at the corner grocery store and that it was cheap and flavorful. They assumed that food production was a simple matter and so science had no need to meddle in the affairs of the farm. Better to confine science to the work of engineering and medicine, where it was needed. With these assumptions in place, few Americans were interested in how food got to the grocer's shelf or whether science had anything to do with its quality or abundance.

Throughout the nineteenth century, America moved along the path of becoming a nation of cities and factories. The end of World War II signaled the next step in this journey. By then the United States had gained skyscrapers and automobiles, and the station could only thrive by satisfying the needs of the city as well as those of the farm. But who would lead the station toward a future in which both farmers and urbanites gained something from its research?

7 THREE SQUARE MEALS A DAY

EDMUND SECREST RETIRED IN 1947 AMID WHAT JOURNALIST WALTER LIPPMAN called "the American century." From their factories, Americans had produced the planes and tanks that defeated fascism in Europe and imperialism in Asia. Americans had returned home to raise families and build schools, hospitals, and roads. Victory abroad and prosperity at home did more than content Americans. The two elevated the prestige of business leaders.

And so, when Secrest retired, a scientist would not do in his place. Indeed, if the right man were available, he could become both director of the station and dean of the agricultural college at The Ohio State University, for both positions were vacant. The trustees wanted "a really big man," and in an America of unprecedented affluence, that man could only be a business leader. Charles F. Kettering, president of the Board of Control, echoed the trustees by advocating the appointment of a leader who would bring "prudent management" to the station.

Despite the fact that both the Board of Control and the trustees wanted the same type of person at the top—a business leader—they remained free to retain the tradition of different men at their helms. And in fact neither demanded that the same person be their point man. But as early as 1947, Howard L. Bevis, the university's president, confided to an editor of the *Wooster Daily Record* that he had been seeking closer ties with the station for several years. He now saw his opportunity, if only he could combine dean and director in one man.

Visitors hear about swine research at a field day in the 1950s.

But who? Kettering had in mind the businessman Leo L. Rummell, who had grown up on a farm near Niles, Ohio, in an era when horses still outnumbered tractors. Rummell graduated valedictorian from the local public high school in 1911 and hoped to study law at Ohio State. His father, however, insisted that he learn something "practical," which meant agriculture. Rummell obeyed his father, though he fraternized with pre-law students and, with them, joined the debate team as its first agricultural student. There he befriended John W. Bricker, who would later become a state senator and governor.

In 1915, Rummell graduated with a bachelor's degree in agriculture, and that year Thorne hired him as the station's first editor. While working for Thorne, Rummell earned a master's degree in journalism from Ohio State in 1917 and the next year joined *Ohio Farmer* as an editor. In 1939, Bricker, then governor, appointed Rummell to the station's Board of Control, and in 1946 Rummell became its president. Meanwhile, since 1942, Rummell had served on the boards of directors of Ohio Bell Telephone and Union Stockyards of Cleveland, and since 1944, as director of public relations for the Kroger Company of Cincinnati. In 1945, Bricker also appointed Rummell a trustee of Ohio State. Rummell thereby joined Joseph Brigham and Lucius Wing, a banker, as the only simultaneous members of the Board of Control and the Board of Trustees.

But not everyone was happy to see this businessman as dean and director. Upon his appointment, a group of roughly one hundred farmers protested that the trustees had not selected "the best candidate." In an era when agricultural scientists were researchers rather than publicists, they believed Rummell lacked the research credentials necessary for the job. Much later, Rummell would concede that farmers had not universally praised his appointment. In February 1948, however, Rummell ignored this protest and instead addressed a friendly audience at the Wayne County chapter of the university's alumni club. There he expounded a simple message: the agricultural college existed to educate young men and women to help Ohio farmers produce food more efficiently. At this gathering, he displayed his talent for building consensus by arguing that farmers needed to unite to realize their common objectives of efficient production and distribution. He punctuated this message by requesting "the full support of the agricultural people in all of Ohio's eighty-eight counties."

Despite his critics, Rummell brought considerable talents to the station. A long career that required Rummell to fraternize with corporate heads and the wealthy farmers who dominated the state's agricultural organizations never diminished his belief that less-affluent farmers were the foundation of Ohio's agriculture. In 1950, for example, he opposed the attempt of Donald F. Jones, a prominent corn breeder, to patent hybrid corn varieties because Rummell feared that an increase both in seed prices and the clout of seed companies

Workers and their cat, all unidentified, at a Wayne County dairy. The location is unknown but is probably not the station, which had only a small dairy. The uniforms and the size of the milking crew are typical of a larger facility.

Leo L. Rummell, director, 1948–1960.

would diminish the purchasing power of less-wealthy farmers. At his best, Rummell spoke in vivid and sometimes poetic language that ennobled farming and agricultural science. His statement that "Science is plowed deeply into farming" elegantly reassured farmers of the natural compatibility between science and farming. Most importantly, he recognized his own limitations and, as a non-scientist, permitted the station's senior researchers to set their research agendas. Rummell cultivated an affable and informal image that won him the allegiance of his staff, as is evident from a letter sent from two dairy scientists in January 1952 thanking him for his "open-door policy."

But Rummell was not all charm. Like Thorne he could be combative. He disliked Democratic Governor Frank Lausche, who had defeated his friend John Bricker in the 1948 gubernatorial election. As dean of the College of Agriculture in 1949, Rummell thwarted Lausche's efforts to appoint Lawrence A. Kauffman, animal sciences chair, as chief of the Ohio Department of Agriculture. Rummell cloaked this thrust at Lausche by appealing to Kauffman's contract with the university. In reality, this move was part of a partisan attack in which Rummell invited the Republican gubernatorial candidate to tour the station in July 1950. Rummell went further than the appearance of partiality by endorsing Bricker against Lausche that year.

In contrast to Thorne, however, Rummell did not find himself mired in controversy. What acrimony there was centered on how the station tenured and promoted scientists. Since the late nineteenth century, the research model had come to dominate academe.

To be sure, the Ohio station was separate from Ohio State, and so the station's scientists could claim exemption from the demands of scholarship. But this argument sounded shrill by the time Rummell became director. Because he headed both the College of Agriculture and the station, one could hardly insist that standards of tenure and promotion be dependent on whether a scientist worked in Columbus or in Wooster.

Equally important, during the twentieth century, scientists with doctorate degrees came to dominate the station. When Glen Stringfield came to Wooster in 1924, he was among the last of his kind—a station scientist with only a bachelor's degree. Two years later, William Krauss, who had a doctorate degree in dairy science, joined the station. In 1930, Leonard Alexander arrived in town with his doctoral research underway; four years later he had his sheepskin. Krauss, Alexander, and others brought with them the research model, and under Rummell, they determined to make this model the norm. After all, Ohio State had required the doctorate degree and publications of its faculty since 1922. Why should the station wait? In October 1953, Alexander, Orville G. Bentley, an animal scientist who would rise to become assistant secretary of science and education in the U.S. Department of Agriculture in 1982, and others requested that the board formally adopt Ohio State's tenure system. Alexander meanwhile solicited Rummell's endorsement. Rummell was evasive, though the fact that Rummell had unified corresponding departments at the university and at the station was enough to sustain Alexander's hope that Rummell would support the scientists' cause.

But in November 1953, station foresters denounced the proposal. One forester dismissed the doctorate degree as "a union card" and maintained that a researcher's accomplishments "irrespective of his having a Ph.D." should constitute the only criteria for tenure. Three others joined him, asserting that "academic degrees … should not automatically supersede all the other qualities" in the granting of tenure. Service to both the station and farmers, they argued, should earn a scientist tenure. The foresters' opposition was understandable since none of the four possessed doctorate degrees. At the time, forestry was a discipline whose practitioners regarded the master's degree as the terminal degree. Despite the foresters' opposition, they could not resurrect a bygone era when generalists typified station personnel. At its January 1954 meeting, the Board of Control sided with Alexander and Bentley by mandating that no scientist could gain tenure at the station without possessing a doctorate degree.

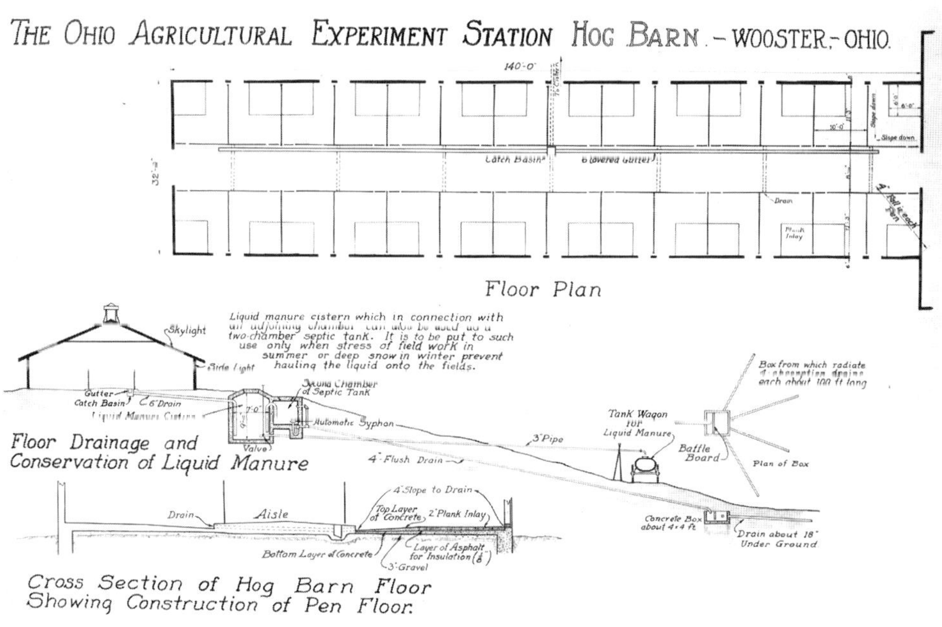

The floor plan of the station hog barn built at Wooster in 1951.

Pulitzer Prize-winning author Louis Bromfield at a 1948 field day at the station. One of his pet boxers lies at his feet. From his Malabar Farm in Richland County, about forty miles west of Wooster, Bromfield wrote *Pleasant Valley*, *Malabar Farm*, and other books about farming and conservation. Like Charles Thorne, he had a passion for experimentation and for restoring soil fertility. Today his farm is an Ohio state park and his books continue to be read.

In borrowing the research model from Columbus, the station was becoming more urban. Perhaps no one appreciated this fact more than Rummell, who was a new kind of director. Although he was born on a farm, by adulthood he was an urbanite. He had made his money at the Kroger Company and had served on the boards of the Ohio Bell Telephone Company and the Union Stockyards. He embraced corporate efficiency and would apply it by consolidating the station to a degree that Thorne could not have imagined.

Rummell knew that Ohio was no longer the land of farms that Thorne had known. To survive, the station would need to prove its utility to urbanites. In 1959, Rummell defended the station's need for more money before the Ohio Finance Committee by framing his argument in terms that urban consumers, rather than rural producers, would appreciate. The station had kept food prices low by ensuring bountiful harvests and kept children healthy by testing vitamins and antibiotics. It had freed most Ohioans from farming so they could live and play in suburbia.

As a businessman, Rummell knew the value of cutting costs and found the substations an inviting target. They had served Thorne too well, for his success in restoring soil fertility now made them superfluous. During the winter of 1948, Rummell toured the substations but found little to admire. One had only rabbits in the horse barn, and at another, the manager had ceased experiments in 1933 so he could raise crops for his own profit. At a third, the four employees took turns caring for their own livestock. At a fourth, Rummell found the manager still asleep at 9:00 A.M. on the day of his visit. At still another, employees had bought five potato diggers though they planted only three acres in potatoes.

Freeman Howlett, horticulture scientist and later chairman of the Department of Horticulture, with an apple tree. The station released the new apple cultivars Franklin in 1937, Melrose in 1944, Ruby in 1953, Holiday in 1964, and Holly in 1970.

Such waste was unimaginable in Thorne's day, and Rummell was determined not to tolerate it now. With approval from the Board of Control and the legislature, Rummell sold ten of the fourteen substations. He replaced them with a fifteen-acre substation in Huron County in 1948, one of 247 acres in Wood County in 1951, and a 275-acre substation in Brown County in 1956. These substations conducted research on, respectively, vegetable culture on muck soils, field-crop production on heavy clay soils, and burley tobacco breeding. By consolidating the substations, Rummell reduced the station's landholdings from the 4,237 acres he inherited in 1948 to 2,755 acres in 1956.

Rummell also found the station to be an eyesore. In January 1948, he mandated that scientists and technicians discard all debris within ten days. That spring, he ordered the maintenance staff to landscape the grounds, paint all wooden buildings, and repair dilapidated fences. By summer, townspeople and the local media, including the *Wooster Daily Record,* began to praise Rummell for renovating the station.

But Rummell was not done yet. He built a new powerhouse in 1948 to replace the one Thorne erected in 1909. In 1950, Rummell constructed the horticulture building and in 1955 the maintenance building. The following year he purchased sixty-eight acres on the southern edge of the station for fruit culture. In 1957, he completed the agronomy and forestry building, named Williams Hall in honor of Carlos Williams, and in the following year he completed the Agricultural Engineering Building.

Yet one can exaggerate Rummell's influence. He spent most of his time at

Joseph B. Polivka, an entomologist who championed the use of DDT for pest control. Rachel Carson's 1962 *Silent Spring* led to the banning of DDT in the United States because it harmed wildlife, especially birds. Since then, station scientists have continued to work to make pesticides and application methods safer.

his alma mater in Columbus and was dean first and director second. To do the daily work of administration, Rummell appointed William Krauss in 1948 as the first associate director. Krauss had joined the station in 1926 as a dairy scientist. During the next decade he led a team of scientists in demonstrating the importance of vitamins A and D in human nutrition. The result of this work, vitamin-fortified milk, has thereafter been a dietary staple.

Krauss, like Viola and Charles Thorne, Leonard Alexander, and so many others, viewed the station as a family rather than merely as a workplace. For Krauss, Christmas symbolized the unity of this family. In 1948, he began a twenty-one year tradition of making a speech to the scientists and staff who attended the annual Christmas party. The Christmas message, as it became known, emphasized their unity as people whose common values transcended differences of ethnicity and gender. He spoke of his hope that nuclear war would never engulf the world and that science would continue to bring food and medicine to people around the globe.

Even Krauss had to admit that Rummell, from behind the dean's desk in Columbus, revitalized the station through refurbishment and a building campaign. Rummell also obtained more state money for the station during his eleven-year tenure than the station received between the years 1882 and 1932. Rummell won larger appropriations by stressing that agricultural research benefited all Ohioans by keeping food prices low. Food prices were low because American farmers were the world's envy—their granaries overflowed with wheat and their barns were crowded with cattle. The station existed to serve farmers by helping them to produce even more food and, in doing so, to ensure that all Americans would enjoy three square meals a day, as Rummell was fond of saying. Food surplus, he believed, was part of every American's birthright.

Rummell was confident that the station could best help farmers produce more food by helping them tailor production to demand. Consumers did not want to eat just any cut of pork; as early as 1936, the *Columbus Evening Dispatch* recognized that Americans wanted leaner cuts. To meet this demand, stockmen needed to raise lean hogs, and Rummell wanted the station to help them do it. In 1950, Rummell built the Swine Evaluation Center and put it under the joint management of the university and the station. Within a decade, the center had bred hogs that were 40 percent leaner than the average hog of 1950.

Americans wanted not only leaner meat, they wanted more of it. And the post-war prosperity allowed them to buy more meat than they had been able to afford during the Great Depression. Consequently, stockmen needed to produce more meat, and one way to do so was by reducing the time required to raise cattle and hogs to market weight. Here again, the station had the answer. During the 1950s, Earl W. Klosterman led a team of animal scientists in demonstrating that beef cattle, hogs, and turkeys fed rations with steroids and antibiotics reached market weight in only eighty percent of the time required by livestock and poultry on unsupplemented rations.

The barn was not the only place where the station applied the wonders of modern chemistry. In 1948, entomologist Joseph B. Polivka turned a campus barn into a tourist attraction by spraying the barn's sides and the ground with DDT. Astonished visitors strolled through the barn without seeing or hearing a fly, and even *Ohio Farmer* was impressed enough to feature the barn in its pages. Polivka liked the toxin so well that he also sprayed it on potato and tomato plants and on peach and apple trees. By 1956, however, Polivka had

Harmon Runnels and Roy Thomas, plant pathologists, catalog diseased plant tissue in the 1930s.

switched to chlordane because it was more potent than DDT. Chlordane was especially deadly to Japanese beetle larvae, and he recommended that homeowners spray their lawns with it annually. (Because of concern about damage to the environment and harm to human health, the U.S. Environmental Protection Agency banned all uses of chlordane in 1988. DDT use in the United States was banned in 1973). As Secrest had done, Polivka was using science to benefit an urban middle class as well as farmers.

Chemistry shaped the station in other ways during Rummell's tenure. Agronomists taught farmers to inject anhydrous ammonia into the soil and to spray a new class of herbicide on their fields, one that killed broadleaf weeds but spared corn. Thanks to the station, farmers could avoid hours of drudgery by applying a few chemicals to their land.

Unfortunately, the control of crop diseases was not so easy. In 1948, a fungal disease swept through soybean fields in western Ohio and Indiana, leaving the plants to rot in its wake. In 1954, A.F. "Fritz" Schmitthenner, a plant pathologist at the station, and Albert J. Suhovecky, a graduate student at Ohio State, identified the disease as *Phytophthora* root rot, and three years later Schmitthenner found a gene resistant to the pathogen. During the next three decades he helped to breed thirteen *Phytophthora*-resistant soybean varieties, which farmers have grown on as much as half the soybean acreage in Ohio. Thanks in part to Schmitthenner, Ohio farmers have, since 1987, grown soybeans on more acres than any other crop.

When Rummell retired in 1959 the station had surely exceeded even Thorne's grandest hopes. Because of its work, Americans ate lean meat and as much of it as they could afford. With the spray from a nozzle or the push of a button, homeowners could banish Japanese beetles from their yards, and farmers could minimize weeds in their fields. Farmers could raise soybeans without fear of catastrophic losses. What more could anyone have expected from the station?

8 RESEARCH, EDUCATION, AND EXTENSION

SUCH A QUESTION WOULD HAVE ANNOYED RUMMELL'S SUCCESSOR, ROY MILTON Kottman. Kottman's life proved that more could always be done. He was born in 1917 on a 240-acre farm in Thornton, Iowa, a town of roughly two hundred people where, until the early 1930s, children rode a horse-drawn bus to a one-room schoolhouse. After graduating as valedictorian from high school in 1934, Kottman worked as a farmhand for $1.00 per day, then as a gas station attendant. Remarkably, Kottman worked his way up to manage and then own a small chain of gas stations by 1936. His real interest was agricultural science however, and in 1937, he sold the chain and entered Iowa State College from where he graduated with a bachelor's degree in animal husbandry in 1941. After Pearl Harbor, Kottman enlisted in the U.S. Army and was commissioned a second lieutenant. During the war he taught courses at West Point, an experience that sparked his desire to become an academic. After his discharge from active duty in 1945, Kottman returned to Iowa State College, now Iowa State University, where he earned a doctorate degree in animal breeding in 1952.

Roy Kottman, director, 1960–1982.

Roy Kottman and Associate Director William Krauss in a publicity photo for the 1965 name change—from the Ohio Agricultural Experiment Station to the Ohio Agricultural Research and Development Center. Krauss wields a crowbar.

In 1958, Kottman became director of the West Virginia Agricultural Experiment Station and dean of the College of Agriculture, where one scientist recalled him as "a born leader." There, Kottman, who stood six feet three inches, established a reputation as a powerful and persuasive advocate of agricultural production and one who could extract money from the state legislature. In 1959, he sensed his opportunity when Rummell announced his intention to retire the following year both as dean and director. Kottman wanted the challenge of serving the farmers of a state both larger and wealthier than West Virginia. But he also anticipated that the agricultural deanship would provide opportunities for advancement to the presidency of Ohio State, an ambition he never realized. John Bricker, a former Ohio governor, remarked that the trustees hired Kottman because, "He looked honest and straight, and he was a good-looking man." This explanation is more credible than it appears because Kottman's charisma also impressed Novice G. Fawcett, president of Ohio State, who was the most influential person on the search committee. In 1960, Kottman became dean and director, and two years later the trustees named him chief of the Cooperative Extension Service, making him the only person to hold all three positions.

Kottman thereby became the most powerful figure in Ohio agriculture. He may even have surpassed Thorne in his influence. Certainly he tried to outdo Thorne, though Kottman did not strive to do science differently than Thorne had done. Kottman was a traditional production-oriented scientist. He wanted to do more of the research that the station had long done well. And to emphasize his commitment to research, Kottman, at Governor James A. Rhodes' suggestion and with approval from the Ohio General Assembly, changed the station's name in 1965 to the Ohio Agricultural Research and Development Center.

He would have nothing less than the world's finest research institution, he assured Ohioans. This goal required money, and Kottman was better than anyone, even Thorne, in getting it. He had come to Ohio with a reputation as a fund-raiser and proved he deserved this acclaim by giving 250 speeches a year to audiences in every corner and crossroads in the state. Anyone who would listen heard his promises to overflow granaries with wheat and to overflow pastures with cattle. He spoke of a day when Ohio's fields would yield seventy bushels per acre of soybeans and twice that of corn. Farming had made America the greatest power on earth, he boasted, and this grandeur depended on more money for agricultural science.

To be sure this optimism was pollyannaish, but it was part of his magic. Kottman was Ohio's version of Lyndon Baines Johnson; they both had the ability to awe a listener. Kottman would engulf the hand offered to him in a firm squeeze and, while pumping it, would draw close to his listener. He would twist his head, hunch his shoulders, and bend his knees so that he could look the listener squarely in the eyes. Kottman would then draw the listener still closer by placing his hands on the listener's shoulders. All the while Kottman would be talking about hogs and wheat and grapes. Everyone in Ohio, no matter how powerful or self-assured, crumbled under this treatment. One legislator joked that Kottman would have run for governor had it not been a demotion. Few laughed, for it was true.

Maize dwarf mosaic virus appeared in the Ohio River valley in 1962. Station agronomists responded to the threat by establishing a two-acre farm in Portsmouth to test corn inbreds and hybrids for resistance.

Lansing E. Williams, a plant pathologist, with corn seedlings in a greenhouse. He was one of the discoverers of maize dwarf mosaic virus. The photograph was taken in 1965 or 1966.

The year Kottman arrived, the center got nearly $3.2 million from the legislature and just under $800,000 from Congress. By 1979, OARDC collected $10.9 million from the state and $4 million from Washington. No one but Kottman—not even Thorne—could have believed the center capable of gaining such wealth. The General Assembly had always, Thorne had complained, been tight-fisted. Secrest had not even tried to increase the budget because he was sure it could not be done.

Kottman's success brought him attention from Senator John Glenn, who nominated him for Secretary of Agriculture during the Ford and Carter administrations. But other men got the job, and Washington's loss was the center's gain. Another man might have been overwhelmed by the exponential increase in funds. But Kottman knew where every dollar went, and he shifted money with the skill of a general repositioning troops in the heat of battle. He bought the center an electron microscope with funds from the Department of Home Economics at Ohio State and made good the loss by getting the department its money back from the legislature for the next fiscal year.

The image of a battlefield commander is apt for Kottman because he remained a military man in everything he did. After returning home from World War II, he stayed in the Army Reserves, rising to the rank of colonel. At Wooster, an entourage of administrators trailed him as though he were Dwight Eisenhower or Douglas MacArthur. He parted his army of lieutenants like the Red Sea as he went from building to building for a white-glove inspection. And these tours were not just for show. The discovery of dust was a serious matter because it confirmed to him that staff had become complacent. How could scientists do good work in such environs?

But many scientists were irked. They were scholars who were at ease in a collegial setting. They chafed at the rigidity of inspections. Others joked that Kottman was trying to turn the center into a Prussian military academy.

Kottman was aware of these criticisms but they did not bother him. He was too busy helping farmers to be worried about a few grumblings. In 1962, farmers near Portsmouth, Ohio, reported severely stunted corn plants in several cornfields. The next year corn was stunted on 15,000 acres in twelve counties along the Ohio and Scioto rivers, and losses were as high as ninety percent. That year, Leonard Alexander joined plant pathologist Lansing Williams

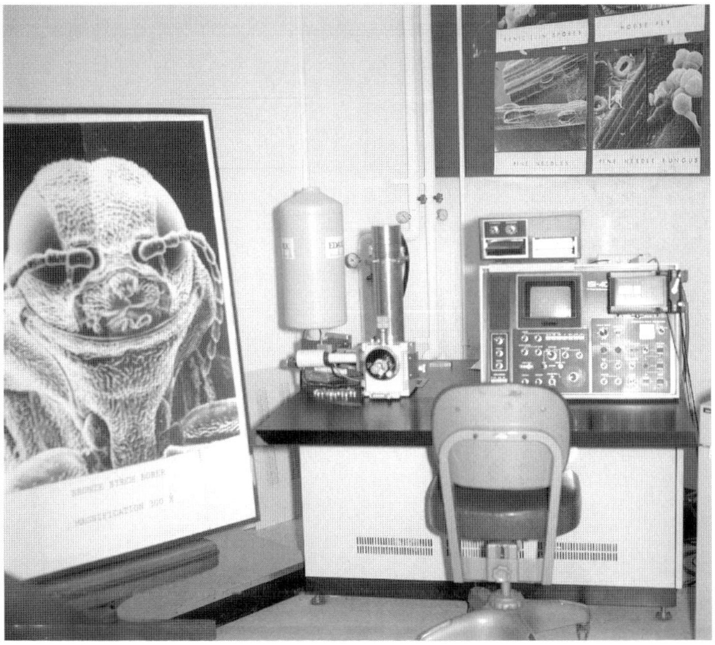

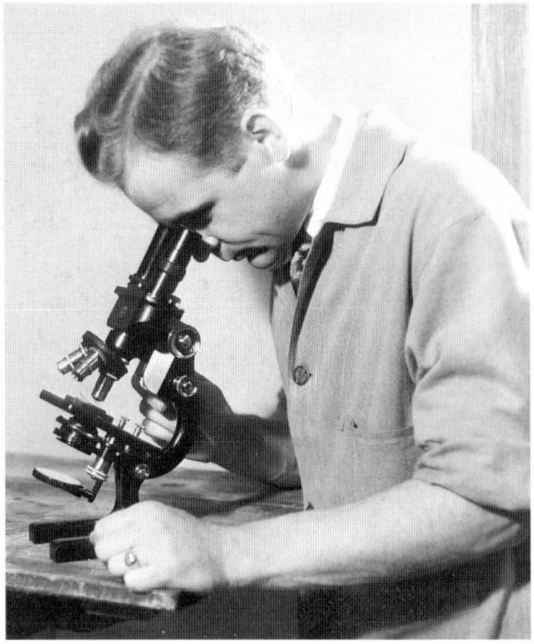

At right is a standard microscope with a magnification of a few hundred times. At left are several examples showing the magnification power of an electron microscope. The photograph at left was taken in the Electron Microscope Laboratory, which is now the Molecular and Cellular Imaging Center.

in identifying the pathogen, maize dwarf mosaic virus, and helped identify the carrier as the corn-leaf aphid. The mission now was to control the disease. The center owned no land near Portsmouth, so in spring 1964, Kottman rented a two-acre field southwest of the city along the Ohio River. There, agronomists William R. Findley, Ellwood J. Dollinger, and Lansing Williams screened hundreds of inbreds and hybrids amid an epidemic of maize dwarf mosaic virus.

That summer Ohio corn growers lost ten percent of their crop, or five million bushels worth $5,850,000, to the virus. The epidemic engulfed seventy-nine of Ohio's eighty-eight counties with the heaviest losses occurring in the southern third of the state, where some farmers lost as much as ninety-five percent of their corn. That year, the virus damaged more corn in Ohio than anywhere else in the severest outbreak of a corn disease until the southern corn leaf blight of 1970.

Thanks to the station, the epidemic did not recur. In 1964, Findley, Dollinger, and Williams identified resistant hybrids and inbreds, including Glen Stringfield's Oh07. In 1965, corn growers planted hybrids that contained Oh07 throughout southern Ohio, and Kottman, who brought journalists and state legislators to healthy corn fields in Portsmouth, credited the hybrids with decreasing the state's losses to the virus from five million bushels in 1964 to 2.6 million bushels. In areas where farmers had planted resistant hybrids, losses were negligible. Only where farmers had not planted these hybrids were losses again severe. As late as 1972, farmers in southern Ohio continued to plant resistant hybrids.

But the center's work against corn viruses was not done. In 1971, a team of scientists, including plant pathologists Oscar Bradfute, Raymond Louie, and Donald Gordon, identified a second virus, maize chlorotic dwarf virus. Entomologist Lowell R. "Skip" Nault discovered its leafhopper vector and the unusual manner in which the virus was transmitted. In 1974 and 1976, Findley bred three hybrids with moderate resistance to both viruses. Such success prompted Kottman and Williams to argue that the center had saved corn culture in southern Ohio.

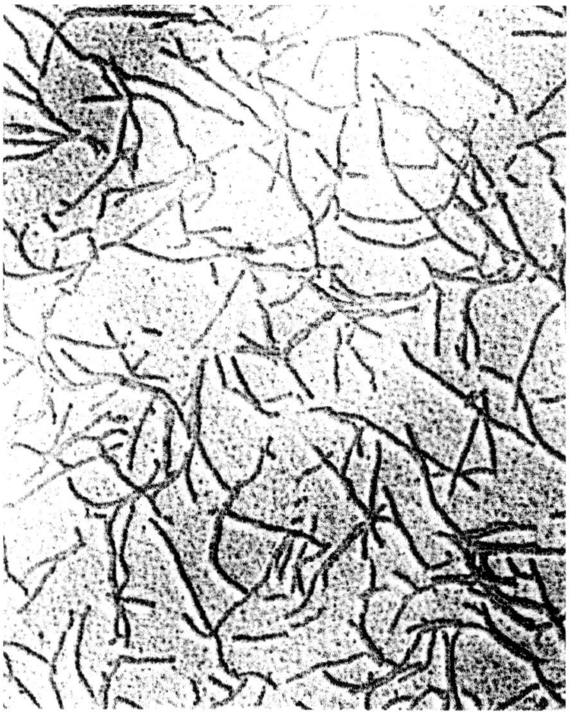

Maize dwarf mosaic virus as seen through an electron microscope.

Other successes followed the struggle against corn viruses. Two U.S. Department of Agriculture entomologists at Wooster—T.L. Ladd and Michael Klein—working with USDA chemists in Florida, found sexual attraction to be a weapon against insects. In 1973, the team isolated a pheromone, a sexual scent, from female Japanese beetles. By synthesizing the pheromone in the laboratory and dabbing it on a trap, the team was able to attract and kill thousands of male beetles. The *Bag-a-Bug* that retails in garden centers today is the result of their work.

Equally important, this work opened up a new era in entomology research by OARDC scientists. Since Thorne's day, entomologists had sprayed increasingly lethal chemicals on insects in hopes of eradicating them. Joseph Houser had bombed them from the air and Joseph Polivka had blanketed them with DDT. But Ladd and his colleagues found a non-toxic biological control for insects. They had satisfied the demand of many Americans for fewer chemicals in order to keep groundwater, lakes, and streams pure.

Corn breeder William R. Findley covers the stigmas of corn plants to keep stray pollen from fertilizing them (circa 1965).

These Americans had, since the publication of Rachel Carson's

Silent Spring in 1962, coalesced into an environmental movement. Chemicals that were thought to be safe were being questioned. People pointed with chagrin to fertilizers seeping into groundwater, insecticides killing robins, and fungicides on fresh fruit. They wanted farmers to quit using DDT and, in doing so, to jump on the ecological bandwagon.

This movement had links to the conservation movement of the late nineteenth and early twentieth centuries, which included Theodore Roosevelt, Gifford Pinchot, and Edmund Secrest. But while the earlier effort drew strength from an urban

The black-faced leafhopper, the vector of maize chlorotic dwarf virus in corn.

Two photographs showing changes in pesticide application over time. Above, from the protection of an enclosed cab, a worker applies a spray in measured amounts. Below, insecticide is sprayed from a hand-held wand with a great deal of drift and little protection for the workers or the horse.

Starting in 1971, center scientists conducted the first U.S. studies of large round bales, shown here wrapped in plastic. Their research, coupled with the development of new equipment by U.S. companies, revolutionized the harvesting, handling, storage, and feeding of hay. Today large round bales are a common sight in Ohio, in other states, and around the world. The photograph was taken at a field day in 1998.

middle class, this new movement was as suburban as the station wagon. It attracted teenagers in tennis shoes who wanted to create and enjoy a cleaner environment. It attracted housewives who wanted to be a part of something other than Girl Scouts. Engineers and lawyers showed their solidarity by bicycling to work. People began to recycle cans.

Kottman was not born in a suburb, but he was no less sensitive to the movement, which he saw as a way to expand the station. In 1962, Congress passed the McIntire-Stennis Act, which, beginning in 1964, appropriated at least $10,000 to each experiment station for forestry research. In 1965, Kottman used this money to establish substations of 121 acres in Fairfield County and 2,093 acres in Noble County. The former, with both arable land and forest, enabled scientists to observe how modern, mechanized farming affects woodland wildlife. The latter, which was larger than any other substation and larger even than the headquarters at Wooster, contained the most deeply strip-mined land in the state and therefore was ideally suited for reforestation research. In 1971, Kottman established a 227-acre substation in Coshocton County to initiate a program in forest tree nutrition and genetics. Two years later Kottman joined with the Bureau of Forestry to manage a 1,047-acre forest preserve and farm, also in Coshocton County. Here, agricultural engineers attempted to determine water runoff rates as well as to measure insecticide, herbicide, and fertilizer contamination of groundwater under different farming methods.

The Eastern Ohio Resource Development Center, established in 1965 at Caldwell in Noble County. The facility is used for research on strip mine reclamation and on forage, beef, sheep, and timber production.

In addition to adopting the sentiments of the ecological movement, Kottman embraced suburbanites by amplifying Rummell's message that the center had kept food prices low for consumers. Kottman also wanted to appeal to city dwellers' desire for beauty amid concrete and high-rises. To this end, in 1966, he supported a program of breeding trees with a broad leaf canopy to shade city streets.

Kottman supplemented the three environmental substations with two substations to serve traditional agricultural interests in Ohio. In 1968, he established the Jackson Branch in Jackson County to determine the best forage grasses and legumes for southeastern Ohio. Nine years later, Kottman established the Vegetable Crops Branch in Sandusky County to test new vegetable varieties, particularly those of field tomatoes. These five substations brought the center's landholdings to nearly 7,500 acres, or nearly triple the 2,755 acres Kottman inherited from Rummell.

Kottman was no less energetic a builder at the Wooster campus. In 1965, Gerlaugh Hall; hog, poultry, and dairy complexes; and a building to house farm machinery, implements, fertilizers, and seeds were built. Two years later, Fisher Auditorium, with a library and a one-thousand-seat auditorium, was finished. In 1972, the plant pathology building, named Selby Hall to honor OARDC plant pathologist Augustine Selby, was completed, and in 1978 a new dairy center was built, which Kottman named in honor of William Krauss.

Although Kottman had expanded the center by appealing to suburbanites, he was not always comfortable in the suburbs. During the 1960s, young men and women from the cities and suburbs flooded Ohio State and other campuses. They wore sandals and tie-dyed clothing. Men sported beards and long hair and gave peace signs in greeting. They marched against the Vietnam War. Some male students refused to register for selective service or burned their draft cards. At the University of California in Berkeley students stormed the administration building, and at Kent State University they burned down the ROTC building. At Kent State the Ohio National Guard shot and killed four students, and at Jackson State University police killed two others.

From the dean's desk, Kottman admitted that he did not understand these suburban students. In contrast to them, he was clean-shaven and his hair was closely cropped. Indeed, so great was his dislike of the hippie image that Kottman did not like to hire anyone with a beard. The multicolored clothes of these students contrasted with Kottman's business attire. The joke was that he owned 365 long-sleeved

Garth A. Cahoon, a grape researcher in the Department of Horticulture. Starting in 1963, Cahoon and James F. Gallander, a wine-production researcher in the department, teamed up to resurrect Ohio's grape-wine industry. The photograph was taken in 1988.

Shade tree evaluation plots in Secrest Arboretum in the 1970s.

white dress shirts. And like his confidant, Governor James A. Rhodes, Dean Kottman viewed the students' protest against the Vietnam War as disloyal. Finally, the equal opportunity movement that drew strength from the suburbs had bypassed the center, which remained overwhelmingly white and male.

In a sense the suburb was too small to contain Kottman. He believed that the center should benefit the kids who did not come from the suburbs and who were unlikely to go to college. These kids needed a two-year technical education after high school, one that would give them practical skills. With them in mind, Kottman persuaded the legislature in 1970 to create The Ohio State University Agricultural Technical Institute, which was built in 1972–1973. Because this institute would be a partner in the center's mission to serve all Ohioans, Kottman allocated it land just south of the main part of campus. Perhaps ATI today fulfills as nearly as is possible the land-grant mission of providing students with a practical agricultural education. Joseph Brigham would be proud of it.

Jack Vimmerstedt, a forester in the School of Natural Resources, overlooks a study site on a reclaimed strip mine.

If the suburb was too small for Kottman, so was just the United States. Kottman's reputation as an advocate of ever more bountiful harvests reached India, where officials in 1966 invited him to establish a national wheat-breeding center. That year he and William Krauss flew to New Delhi, where they guided Indian agronomists in creating a program to breed semidwarf wheat at the University of New Delhi. The intent of the mission was noble but the results did not match Kottman's hopes. The program seldom had the money that Ohio and Indian agronomists believed it needed. Equally problematic was that Indian farming practices remained primitive. Even if farmers could afford a tractor, few would buy one because parts were seldom available to fix it.

India was not the only country to welcome input from the center. Livestock breeders visited cattle ranches in Argentina. Soybean breeder Richard L. Cooper collected soybean plants in China. Frederick E. Deatherage, an animal scientist, showed Brazilian beef processors that they could slow bacterial growth in meat by injecting antibiotics into the vascular system of each cut.

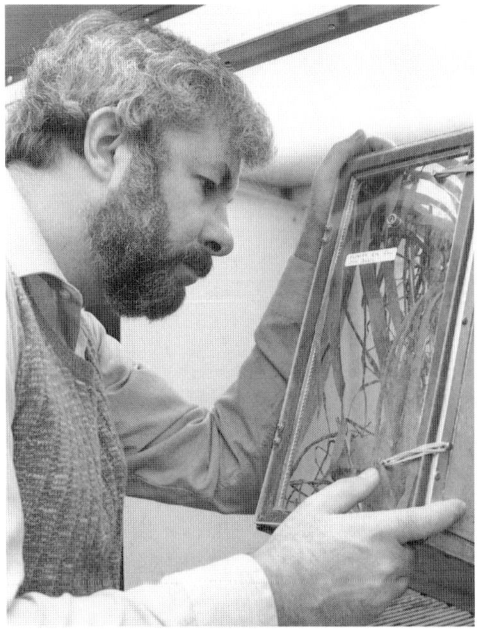

Lowell R. "Skip" Nault (1966).

Perhaps most spectacular was the work of Lowell R. "Skip" Nault, who joined the center in 1966 as an entomologist and, four years later, gained an appointment in the Department of Plant Pathology in addition to his duties in entomology. During his career Nault has become an international expert on leafhoppers, coauthoring *The Leafhoppers and Planthoppers*. He has been an associate editor of *Phytopathology*, has served on the editorial board of the *Journal of Economic Entomology*, and has thrice chaired the publications council of the Entomological Society of America. He is a fellow of the American Association for the Advancement of Science, the American Phytopathological Society and the Entomological Society of America. In 1991, he served as president of the Entomological Society of America. Since 1995, Nault has been the center's associate director; he was interim director from December 1998 to September 1999. (Steven A. Slack became director in October 1999).

Nault's work against corn viruses is only a small but important aspect of his career. In an attempt to locate a source of resistance to several of these corn viruses, Nault went to Mexico in November 1979 in search of a newly found teosinte species. He reasoned that since teosinte, an ancient ancestor of corn, had originated in the Tehuacan Valley of Mexico, he might find this teosinte to have exceptional properties. Having survived so long as a species, it would likely be resistant to many diseases. In a mountain valley in west-central Mexico, Nault collected seeds from this hitherto unknown teosinte, *Zea diploperennis*, and brought them back to Ohio. There he found the plant was immune to many of Ohio's corn viruses. By crossing teosinte with corn, this new source of genetic material could protect Ohio corn, reinforcing the view of many that

Dan Houston, forester in the School of Natural Resources, in an OARDC seed orchard. He continued research by Howard Kriebel of the center that led to "super-sweet" sugar maples and faster-growing white pines (1992).

the discovery of this teosinte was "the most important botanical find of the century."

With Nault's triumph, OARDC added its imprint in Mexico, Argentina, India, and China, just as Neil Armstrong left his footprints on the moon. Kottman had added branch stations throughout Ohio and brought the center's mission in line with the environmental movement. Scientists had beaten back corn viruses and used the scent of sexual attraction to lure Japanese beetles to their death. The center had coaxed more corn from Ohio's farms and given shade to long stretches of asphalt. Under Kottman, OARDC had grown beyond Thorne's grandest hopes. The center, however, could continue growing only with more money from the legislature. But would the legislature, so parsimonious by tradition, pour even more money into the center?

In this photograph taken in 1999, Skip Nault poses with William E. "Brit" Kirwan, president of Ohio State, who has just presented Nault with a University Distinguished Scholar Award, symbolized by an apple. In addition to his stellar research career, Nault began serving as associate director in 1995 and was interim director during 1998–99. The apple is a Melrose, a variety released by the station in 1944.

9 HARD TIMES AGAIN

ALTHOUGH MANY PEOPLE LIKE A FAIRY TALE WITH A HAPPY ENDING, OTHER PEOPLE are drawn to the tragedy, perhaps because it is cathartic. The reader grieves for Oedipus, Hamlet, and Willie Loman as these characters spiral down toward destruction.

In its own way, OARDC fell from great heights during Kottman's last years as director. Skip Nault's discovery of the new teosinte genetic potential marked the center's apex under Kottman, for 1979 marked the beginning of financial woes. Between 1979 and 1981, the legislature, desperate to cut expenditures, reduced its appropriations to OARDC by seven percent. In raw numbers, between 1979 and 1981, state appropriations dipped from $10.9 million to $10.8 million, and again between 1982 and 1983 from $11.7 million to $11.5 million. In response to these cuts, Kottman froze all research budgets, hirings, and equipment purchases.

By April 1981 the tight budget led Donald T. Gordon, a plant pathologist with the center, to complain in *Ohio Farmer* that he and his colleagues who were fighting corn diseases could no longer afford to maintain equipment, let alone upgrade it. The financial plight was particularly acute since four of the eight scientists who worked on corn diseases were from the U.S. Department of Agriculture, which mandated that the scientists receive annual raises. With a constant budget, the USDA scientists had less money to spend on research in each year between 1979 and 1981. Gordon also conceded that he and his colleagues no longer had enough technicians to perform the numerous assays necessary to identify corn pathogens. He was frank in admitting that research had stultified and that he had seen "a souring of the attitudes of the researchers."

Gordon and others had reason for alarm. Only twice before in the center's 100-year history had appropriations declined. In 1887, the legislature had diminished appropriations from $5,000 to $1,500 in anticipation of Hatch funds. And between 1929 and 1935, appropriations fell from $730,800 to $299,700 as the Great Depression forced every institution to tighten its belt. Some researchers lost jobs and others went on half pay. But in the 1980s, Congress was not waiting in the wings with more money nor was the nation in the grips of a depression.

What went wrong? The trouble was not with OARDC but with the high cost of doing business in Ohio. The New Deal, that avalanche of reforms during the Great Depression, made the federal government the protector of unions, which spread through the automobile, steel, and rubber factories that rimmed the Great Lakes. During World War II and for thirty years after, Americans and the rest of the world bought American goods and American business could afford to pay high wages. But the oil embargo of 1973–1974 and competition from Japan and Germany forced American businesses to cut costs. The answer was for business to relocate the union shops of the Great Lakes to the non-union south. This exodus of corporations and workers shrunk Ohio's tax base and forced the legislature to cut its spending. Like other public institutions, OARDC had to make do with less money.

To be sure, the legislature deserved a share of the blame for the center's plight. It had never funded the center as generously as other states had funded their experiment stations. William Lazenby had to make do with between $1,500 and $3,000 per year when every other experiment station had an annual budget of at least $5,000. From Lazenby to Kottman, the center's directors all complained with good reason about the legislature's parsimony. In 1950, Ohio ranked forty-second among the states in terms of money invested per rural family in agricultural research. Although Ohio ranked fifth among states in the proportion of rural residents and sixth in farm income, the Ohio Department of Agriculture estimated that the legislature allotted only six cents per rural family to the station in 1950. This level of funding meant that it received less than one percent of state appropriations that year. Among Corn Belt stations, Ohio's funding ranked behind those of Illinois, Iowa, Wisconsin, Michigan, Minnesota, and Indiana. In 1979, the Ohio legislature invested only $0.86 per person in the center. In contrast, Minnesota invested $4.14 per person, Indiana $1.31, and Michigan $0.98 in their experiment stations.

A spartan budget plunged OARDC in 1979 into a deficit for the first time in its history. From less than $1 million that year, the deficit gap would widen to $9.4 million in 1992. These deficits were tough for OARDC but at the same time made the center attractive to Ohio State's trustees. The center had, of course, been independent from the university, at least legally, since its founding. But if it were now made a part of Ohio State, the trustees could appropriate money from the College of Agriculture to cover the deficit. The center would then repay these loans not to the College of Agriculture but to the university's general fund. Money that originally could have been spent only on agriculture would become available for other uses. In fact, the bigger OARDC's deficit, the more discretionary money the university would get.

Ohio State, therefore, had a powerful incentive to co-opt the center, something it had wanted ever since Norton Townshend proposed the creation of an experiment station in 1880. The university could argue that it and the center had become increasingly connected. In 1948, Rummell unified common departments at both the university and at the center under a single chair, and since 1962, Kottman had run the university, the center, and the extension service as one unit.

For their part, OARDC administrators and scientists were less thrilled with the prospect of merging center and university. Kottman admitted, however, that by 1981 the Board of Control hoped that the legislature would be more generous to OARDC if it were part of the university than if it remained a separate and smaller institution.

Given these realities, in January 1981 the Board of Control and the Board of Trustees began to discuss unification, and in May 1981 they sought approval from the legislature to merge the two institutions. Two months later the Ohio House of Representatives passed a bill to that effect, and in January 1982 the Ohio Senate concurred. The act modified the law that had created the center in 1882 by replacing the Board of Control with the trustees. In June 1982, Clive W. Donoho, the center's associate director, described the merger as merely a symbolic event by remarking, "We've been one organization all along. It's like a couple that has lived together for 20 years deciding to get married." He stressed that the merger would not "make for any large changes in the center's day-to-day activities."

Stanley Z. Berry in a field of processing tomatoes. A tomato breeder, Berry developed and released a long list of improved processing-tomato varieties. The varieties have been grown on thousands of acres in Ohio and the Midwest. The photograph was taken in 1989.

Clive W. Donoho, director, 1983–1984, and associate director, 1973–1982. He succeeded Roy Kottman as director but retired within a year. He equalled Norton S. Townshend for the shortest tenure of any director.

But not everyone applauded the merger. In April 1981, John K. Flessel, an entomologist, admitted his fear that OARDC would lose its identity as part of a "mega-university" and that research would suffer accordingly. Others viewed the unification not as a merger but as a hostile takeover, and they regarded the trustees as absentee landlords. Perhaps the most widespread criticism was that the university never had the spirit of a land-grant college. Everyone with the center knew that since Ohio State's founding it preferred the city to the farm and a liberal-arts curriculum to agricultural subjects. As early as the 1870s, Joseph Brigham and Charles Thorne had warned farmers that the university had little use for agriculture. Since those days, Ohio State had become more, rather than less, committed to educating doctors, lawyers, and architects. How could it now guide agricultural research after having turned its back on the farm for more than a century?

The unification of center and university ended Kottman's tenure. In 1982 he turned sixty-five. As a university administrator (dean or above), he was obliged to resign his administrative position, and so he retired. He was still vigorous in the way that Thorne was late in life, and Kottman did not want to retire. He had dedicated his life to agricultural science. What would he do now? Despite his reluctance, he did his duty as he had since he was a youngster in Thornton, Iowa.

Kottman's retirement that year marked a watershed—from longer to shorter tenures for the directors. Of the full-time directors, Thorne served thirty-three years and Kottman twenty-two. Williams was at the helm sixteen years and Rummell eleven. Secrest had the briefest tenure at a decade, but he remained with the station as a forester and an administrator from his arrival in 1906 until his death in 1949. During these forty-three years, Secrest remained, like Williams, a quiet, steady supporter of the station. In all, these men averaged more than eighteen years as director. Longevity gave the center stability, and loyalty was second nature to these men. From time to time people disagreed with them; Thorne, for example, fought Berton Carmichael, Grant Case, Dwight Lincoln, Rezin Wasson, and the Ohio State trustees. But no one ever questioned the devotion of these directors to the center.

Their fidelity sank deep roots with the center. Glen Stringfield spent thirty-five years there. Leonard Alexander came to Wooster in 1930 and stayed forty years. Fritz Schmitthenner has beaten Alexander's longevity by five years and counting. William Krauss spent half a century at Wooster and worked with every director from Thorne to Kottman. From top to bottom the center attracted men and women who saw it as a home rather than as a stepping stone on their climb to an administrative post at the U.S. Department of Agriculture or at another university.

When Kottman retired in 1982, loyalty had become old-fashioned. Ford and General Electric shed workers with no more difficulty than a snake gives to casting off its dead skin. Charles Keating plundered the vaults of savings and loans before jettisoning each in turn as he continued to find the next victim.

The directors between 1983 and 1992 were products of the corporate culture in which ambition overrode loyalty. Upon his retirement, Kottman turned over the ship to Clive Donoho and wished him smooth sailing. A horticulturist by trade, Donoho had been associate director since 1973 and Kottman's heir apparent. Donoho spoke of leading OARDC toward the twenty-first century, but his departure in 1984 gave him the shortest stay of the center's full-time directors. Donoho served only as long as Townshend, who had been a part-time director and then only because the Board of Control had not found anyone else. Lazenby, the other part-time director, outlasted Donoho by four years.

Frederick E. Hutchinson, director, 1985–1987. He went on to become vice president for agricultural administration and dean of the College of Agriculture at The Ohio State University, then university provost. He left in 1989 to become president of the University of Maine.

Donoho's successor, Frederick E. Hutchinson, held a doctorate in agronomy from Pennsylvania State University and had been vice president for research and public service at the University of Maine and director of the Maine experiment station. Hutchinson left Maine in 1982 to join the U.S. Agency for International Development, an arm of the State Department, where he distributed federal funds to help poor nations improve farming. He took formal charge of the Ohio center in July 1985 and tried to convince legislators of the value of biotechnology, though he cautioned that results would not be immediate. He combined this sensible approach with an ambition that led him in 1987 to become vice president for agricultural administration at Ohio State, dean of the College of Agriculture, and later the university provost. He left Ohio State in 1989 to return to the University of Maine as its president.

The next man, Kirklyn M. Kerr, would direct OARDC for four years, which beat the combined tenure of Donoho and Hutchinson. Kerr had been a veterinarian at Louisiana State University, and in an era with greater financial stability he might have excelled as director. In 1987, a new administration building was erected and the next year a veterinary science complex was built. In 1989, Kerr added a 293-acre facility in Pike County, the Piketon Research and Extension Center, which underscored the union of center and university. Despite these achievements, the center's deficit frayed Kerr's nerves. As the months wore on, his shirt tail became carelessly tucked and his brow furrowed in anxiety. In 1991, Kerr admitted that he could do nothing more with the budget and resigned.

Kirklyn M. Kerr, director, 1987–1991. He came to the center from Louisiana State University.

James H. Brown, a forester as Secrest had been, followed Kerr as director, but he agreed to only an acting appointment until the trustees found someone else. Brown came to Wooster in 1970 and knew as well as anyone that the deficit was the core of the center's troubles. He admitted that since 1987 the deficit had exceeded $8 million. Because the legislature would not bail out the center, Brown decided to pare expenses until they matched appropriations. Although he did not eliminate the deficit, he slowed its growth so that it reached only $9.4 million by the end of 1992, when it might have been much bigger.

Contruction of Selby Hall at Wooster in 1972. The Department of Plant Pathology and the Molecular and Cellular Imaging Center (formerly the Electron Microscope Laboratory) are housed here.

To be fair to Kottman's four immediate successors, OARDC was difficult to direct during these years. Money was scarce and morale was low. But other men had taken the helm in rough water and steadied the ship. When Thorne became director in 1888, relations between the center and the university were at their nadir. Williams succeeded Thorne with Carmichael still in court, and Secrest guided the center through the Great Depression. Yet all three remained in Wooster, either formally or informally tied to the center, until their deaths.

Amid this procession of directors, fiscal management deteriorated. Whereas Kottman controlled every dollar, Edward Jennings, Ohio State's president, placed the burden of accounting on each department. Because he ordered departments to spend every dollar in each fiscal year as a way of coaxing more money from the legislature, OARDC had no reserve and the habit of fiscal restraint disappeared. One microbiologist was ordered three times to spend the same $25,000 because no one was keeping track of expenditures. On another occasion, the microbiologist was billed $45,000 for fertilizer that he never received and in fact was meant for one of the substations. Meanwhile the substation had the fertilizer but had never been billed for it.

Nor did the changes in leadership between 1983 and 1992 help the center's research. The showpiece was to have been biotechnology. Farmers were about to witness the greatest revolution since they had hitched oxen to the plow 10,000 years ago, boasted Max Lennon, vice president for agricultural administration at Ohio State. In the early 1980s, he and Clive Donoho promised legislators that Ohio's soils would nourish "pomatos," plants with potatos and tomatos. Cattle

James H. Brown, acting director, 1991–1992, and associate director, 1989–1990 and 1993–1995. A forester in the School of Natural Resources, Brown returned to research upon his retirement as associate director. Among his accomplishments: developing the West Virginia balsam fir, or Canaan fir, for use by Ohio Christmas-tree growers.

Visitors inspect turf grass plots at a field in Columbus (1998).

would bear a dozen young per litter. Wheat plants would flourish in saline soils, and corn would endure drought with little yield loss. Cows would give vitamin-rich milk, and pigs would have virtually no fat. These wonders led legislators to expect that some of the $50 million they gave OARDC between 1981 and 1985 would fuel a biotechnology revolution.

But the enthusiasm of Lennon and Donoho was premature. In the early 1980s, scientists were just beginning to match a gene with its trait, and there were millions of genes in the realm of crops and livestock. Even as scientists mapped the genome of each plant and animal, they were only just beginning to be able to remove a gene from one organism and insert it into another. Biotechnology had yet to get off the ground, and so the center's department chairmen often funnelled money for biotechnology into traditional crop and livestock research. What's more, these chairmen were part of the center's old guard. They had crossbred plants and animals the traditional way and saw no point in changing now. Let the rhetoric of biotechnology first prove itself and then ask for money. The triumph that Lennon and Donoho promised fell flat, and by 1986, the legislature wanted nothing more to do with biotechnology.

But traditional programs were not much more successful than biotechnology. For example, during 1988–1990, agronomist Howard N. "Hal" Lafever tried to breed a durum wheat suitable for Ohio, a state with no durum acreage, in the hope of wresting eastern markets from the traditional but more distant durum-producing states of North Dakota and Minnesota. The idea was bold but unrealistic. Even flour millers in Ohio told Lafever that it rained too much in the Buckeye State for growing durum. In 1990, after two fruitless years, Lafever retired when Kerr jettisoned the program to pare expenses.

The durum-breeding program was more than a scientific failure. It was a concession that the agricultural heartland lay west of Ohio, a fact that had been true for more than a century. Lafever had tried to remake Ohio in North Dakota's image. But Ohio was not America's breadbasket, and he was not able to make it so.

Small-grains breeder Howard N. "Hal" Lafever, Department of Agronomy, in a wheat field (1989).

After more than a decade of woes, OARDC had reached its nadir, at least since 1887 when it had only $1,500 from the legislature and Townshend as director. But at that time, Congress had come to the rescue with $15,000 and the Board of Control had hired Thorne. In 1992, no one stood ready to pour money into the center. Ohioans were in no mood to pay more taxes for their schools, let alone for agricultural science. Morale had fallen, directors had stayed just long enough to add a line to their resume, and a few legislators and administrators at Ohio State wondered whether the time had come to sell the center.

10 TOWARD A BRIGHT FUTURE

Charles Dickens might have felt half comfortable at OARDC in 1992. It had been the worst of times. The center had endured 13 years of meager appropriations and had cut corners to avoid falling deeper into debt. Directors had come and gone as fast as the emperors of third-century Rome. At least in 1992 there were no barbarians ready to overrun the center.

But as bad as times had been, they held promise for the future, for 1992 marked a century for the center in Wooster. That July, Roy Kottman returned to Wooster for the Centennial Celebration, and it seemed that he had never been away. He exhorted scientists to redouble their efforts and take pride in

Fisher Auditorium and fountains aglow at dusk. The photograph looks to the east from the main entrance of the Wooster campus.

their center. He shook hands with all the legislators he could find and did not let go until he was certain they understood how much OARDC needed money. E. Gordon Gee, Ohio State's president, echoed Kottman when he asked them for "dollars of reality" rather than platitudes and promises. Charles Thorne's grandson came to the celebration and told people what they already knew—his grandfather had built this center with courage, energy, and perseverance. It would be tragic to let Thorne's lifelong work deteriorate for lack of money. The speeches of Kottman, Gee, and Thorne's grandson brought to life the center's past. But the center's past could lead toward a bright future only with a rebirth of leadership.

Thomas L. Payne, director, 1993–1998. Through BioHio and continuous quality improvement, legislative meetings and staff pizza parties, he rekindled the center's energy and enthusiasm. In 1997, he oversaw the retirement of the center's budget deficit. He left the center at the end of 1998 to become dean of the College of Agriculture, Food, and Natural Resources at the University of Missouri-Columbia.

Among the audience that day was Thomas L. Payne. A native Californian, he had grown up in Bakersfield in the San Joaquin Valley surrounded by some of the world's most fertile farmland. From an early age Payne realized the importance of agriculture to the American economy and to America's global power. America's farms were the breadbasket of the world. Payne carried this knowledge with him to the University of California, the state's land-grant university. At the Santa Barbara campus, Payne chose an unusual combination by majoring in zoology and minoring in psychology. There, his first entomology course kindled his fascination with insect behavior. For graduate school, Payne chose the Riverside campus, which boasts an elite entomology program. There he studied integrated pest management—a way to limit insect damage without relying exclusively on insecticides—on his way to a doctorate degree in entomology.

Upon receiving his sheepskin, Payne became an assistant professor at Texas A&M University. During his eighteen years there he excelled in research, extension, and training graduate and undergraduate students. His colleagues found Payne to be a natural leader in the laboratory, in the classroom, and on the farm.

The two oldest buildings on the center's Wooster campus, built by sons of Frederick Rice, the original owner of much of the campus land. (top) The Barnhart Rice Homestead decorated for the holidays (1998). It was built in 1822 from native sandstone quarried nearby. It is currently being restored. (bottom) The brick Simon Rice House, built in 1822 and used by the center for guests and as the Wooster residence of director Tom Payne (1992).

(left) Daral J. Jackwood of the Food Animal Health Program, who developed a new diagnostic test for infectious bursal disease in chickens that gives results in two days instead of two to four weeks (1997). (right) Linda J. Saif, also of the Food Animal Health Research Program, who among other achievements has pioneered research on rotaviruses in livestock and humans.

Payne's leadership shone at Virginia Polytechnic Institute and State University, where Payne became head of the Department of Entomology. When he arrived in Blacksburg, he saw the same problems that were plaguing other research centers, including Ohio's: high deficit and low morale. To his credit, Payne worked with entomologists at Virginia Tech to create a family spirit where there had been dissension. He also worked to balance the budget.

Before his interview and without fanfare, Payne came to the Centennial Celebration in Wooster as the man who could fix what was broken. The next day he interviewed for the position of director. With his demonstrated ability to manage money and build bridges between people, Payne won the job. He also became associate vice president for agricultural administration and associate dean for research in Ohio State's College of Agriculture (now the College of Food, Agricultural, and Environmental Sciences). In January 1993, Payne moved into the Rice House on campus as the tenth full-time director.

Kottman had suggested that Payne live in Wooster, though his choice to live in the Rice House was all Payne's own. It was a decision that linked Payne to the center's earliest history and thereby gave his tenure a permanence that would have been difficult to achieve any other way. In 1821, Congress gave the land now occupied by the center to Frederick Rice for his service in the Revolutionary War. The next year Rice gave half the land to his son Simon,

Sally A. Miller, Department of Plant Pathology, and Casey W. Hoy, Department of Entomology, pose for a photograph for the OARDC/ATI Speakers Bureau booklet (1995). They work together and with others on integrated pest management of vegetable crops. Through a combination of methods, integrated pest management helps farmers control insects and diseases with minimal pesticide use.

The Horticulture and Crop Science Greenhouses, dedicated at Wooster in 1998. A central computer controls most of their functions. The curved roofs, a new design, are rare on U.S. greenhouses but more common on European ones. They let in more sunlight than angled roofs. The view is from the Research Services Building.

who built a brick farmhouse that is known today simply as the Rice House. It lies just east of the maintenance building and shares the honor of being the oldest building on campus along with the house built by Simon's brother, Barnhart.

David L. Zartman, Department of Dairy Science (now Animal Sciences), and two of his charges at the Mahoning County Experimental and Educational Farm near Canfield, Ohio (1989). At left is a Jersey, at right a Holstein. From 1987 to 1991, Zartman led the Mahoning County Dairy Program, a detailed study of intensive grazing and seasonal breeding. It offered hope and new methods for farmers with small dairy herds.

Whether in the Rice House or in his offices in Wooster or in Columbus, Payne brought a relaxed style to the center—one that he called "personal, proactive, and somewhere between totally democratic and autocratic." Payne broadened the perception of OARDC from being just a place in Wooster to an organization that included faculty and staff on the Columbus and Wooster campuses and at the many branch stations throughout Ohio. The white dress shirt and solid-colored necktie of Kottman's day gave way under Payne to the knit shirt and open collar. He hosted pizza parties at the Rice House, to which scientists, staff, and students flocked at 3:30 P.M. on Fridays. They came in tennis shoes and talked about their children between bites of pepperoni and mozzarella. As further evidence that Payne crafted an informal style, he hosted brown-bag lunches with faculty, staff, and graduate students both in Wooster and in Columbus. Under Payne, OARDC once more was a community of people rather than merely a workplace.

An aerial view of the Wooster campus in 1998, looking east. Madison Avenue is at the bottom of the photograph. The main entrance and exit are at the bottom right.

Payne's style was equally relaxed because, by his own admission, it was democratic. Whereas Roy Kottman was Dr. Kottman, Payne was "Tom," signed with a smiley face in the "o," and he needed no title as a mark of his position. His fresh approach was far removed from the days when Thorne tried to ban off-duty smoking and drinking and Kottman conducted white-glove inspections. In place of this rigidity, Payne shared the responsibility for making decisions with others. He was quick to say that he consulted other administrators, scientists, staff, and graduate students before he reached a decision on most matters. Payne embraced the tenets of continuous quality improvement, or CQI, and by involving everyone in improving the organization he helped scientists, staff, and graduate students to regain pride in it.

Payne's style brought OARDC into the 1990s. Scientists started reading not only the latest articles in their specialty, but also Stephen Covey's *The 7 Habits of Highly Effective People* and his *Principle-Centered Leadership*. CQI linked OARDC to Ford and Honda and to Colin Powell and President Clinton. From factory to laboratory, the lines of decision making moved from vertical to horizontal.

Visitors sensed a new enthusiasm in the workers fitting a huge rectangle of glass into one of the new greenhouses and in the roar of lawn mowers and weed trimmers manicuring the grounds. The Poultry Research Center, the

Sudhir K. Sastry, Department of Agricultural Engineering (now Food, Agricultural, and Biological Engineering), left, and Q. Howard Zhang, Department of Food Science and Technology, in Columbus in a photograph taken for the 1995 OARDC annual report. Sastry and Zhang refined aseptic packaging and pulsed electric field technology to improve the taste of ready-to-eat foods.

Composting Research Center, and the Horticulture and Crop Science Greenhouses were built after Payne rallied support from legislators and industry. Then he joined Bill Anderson, director of ATI, in building Applewood Village, an apartment complex, and planning the Center for Education and Economic Development. Along Madison Avenue, a digital sign welcomed motorists to the Wooster campus in bold, red letters. An older generation of talented scientists, including H. Russell Conrad, Lansing Williams, William Findley, and Fritz Schmitthenner, retired. Assistant and associate professors in their twenties, thirties, and forties—like Lucy Ward, Sally Miller, David Francis, and Sandy Velleman began to fill the corridors. The new researchers seemed too young to have amassed lists of publications that cover several pages of their curriculum vitae. Tom Payne himself was only fifty-two when he became director, and his youthful dynamism matched the vigor of Charles Thorne and Roy Kottman.

With this energy, Payne tackled the deficit. Within months after he took charge of OARDC on January 1, 1993, the deficit had reached $9.4 million. The center had been in the red for more than a decade. Every director since

(from left) Research associates Peggy Lewis and Kathy Gadfield and scientist Linda J. Saif, all of the Food Animal Health Research Program, feed a calf in the program's gnotobiotic facilities. The calf is in a germ-free "bubble," providing a neutral test ground for studying diseases—in this case, winter dysentery (1991).

Kottman had fretted over the deficit but had not been able to do much about it. Kirklyn Kerr admitted that the deficit caused him to resign. James Brown, as acting director, managed to slow its growth, but even he had not been able to shrink it.

From the outset, Payne knew that the deficit was the center's bugaboo. He might have avoided the pain of fiscal belt-tightening had he asked the trustees to cancel the deficit, but he knew OARDC could remain credible only by paying its debts. Payne was left therefore with the hard work of paring expenses. When the Emperor Vespasian wanted to cut costs in first-century Rome, he simply reduced the pay to his troops. But Payne did not have recourse to such a draconian solution. Instead he did not replace retirees. When the legislature further reduced appropriations to OARDC by nineteen percent in fiscal year 1992, Payne kept morale from sinking further by avoiding additional layoffs except for one administrator in his office. Less painful was Ohio State's offer of retirement to faculty, including those with the center, with at least twenty-five years of service. The offer stung only those who wanted to retire early but found themselves too short of years. Outlying facilities likewise shrunk. Payne closed the Robinson Swine Center in Wooster and the Overlook Farm in Carroll and Fairfield counties and reduced operations at the Mahoning County Experimental and Educational Farm.

At the same time, Payne needed to convince legislators that OARDC deserved more money. In doing so, he was quick to say that everyone in the organization helped persuade the legislators through the quality of their work. Bringing OARDC to the attention of legislators required scientists to promote the center to all Ohioans. The extension service had been the vehicle for publicizing the center since Augustine Selby, Charles Thorne, and L.H. Goddard worked to create it at Ohio State more than eighty years ago. Payne wanted the closest possible cooperation between OARDC and Ohio State University Extension. To that end he partnered with OSU Extension Director Keith Smith and the center's assistant to the director for field operations, Jim Trotter, in inviting agents to do research at the outlying branches. In addition, Payne created liaison committees at many of the branches to coordinate programs.

Karl E. Nestor, Department of Animal Sciences, in a Wooster campus turkey barn (1998). His research on leg problems in fast-growing turkeys has helped the poultry industry and has drawn national media attention

Perhaps the most visible method of bringing OARDC to the notice of all Ohioans has been BioHio. In 1994, 1996, and 1998, faculty, staff, and students held three-day open houses on the Wooster campus. From the beginning, BioHio has been a cooperative venture among OARDC, ATI, OSU Extension, and the academic programs of the College of Food, Agricultural, and Environmental Sciences. Visitors watched a man in what looked like a spacesuit tame a swarm of bees. Tours left every 15 minutes with dozens of passengers to survey the OARDC and ATI campuses. During these tours, the eye took in but a few of the thousands of tree specimens in the Secrest Arboretum. Cattle and hogs attracted curious children. Scientists and visitors talked about how much science is doing to feed the world. All this did not cost visitors a penny because admission was free. In 1994, between 10,000 and 15,000 people attended the first BioHio, and attendance grew in 1996 and again in 1998. (BioHio 2000 was not held because of campus construction. BioHio 2001 will be held in Columbus.)

Scenes from the BioHio 96 open house, July 19–21, 1996, at OARDC's Wooster campus and the Agricultural Technical Institute. (top) An African millipede crawls on the head of a Bug Zoo visitor. (bottom) Climbing and clowning on a tractor display. The children are unidentified.

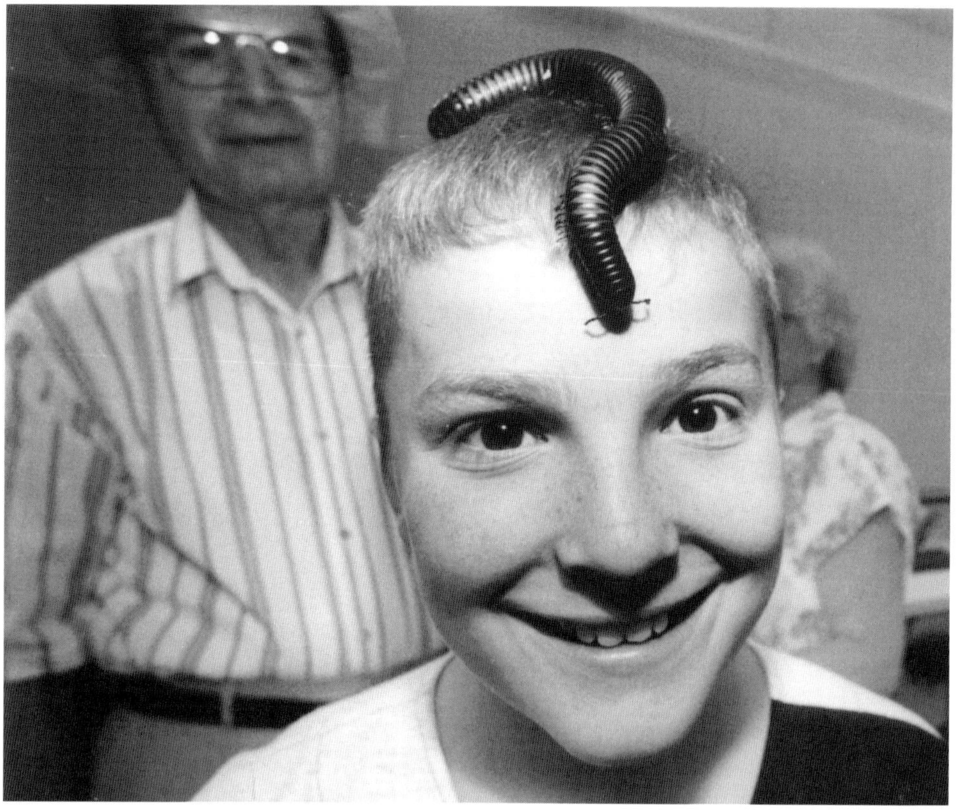

Scenes from BioHio in 1994. At left, petting a cow. At right, one of the thousands of children who received free balloons from the OARDC Staff Council.

In the effort to publicize the center, Payne did his part. He met with State Representative Ron Amstutz and visited State Senator Grace Drake and her legislative assistant, Ted Berry, to remind them of the center's need for more money to continue serving all Ohioans. He invited other legislators and industry leaders to visit the Wooster campus and made sure stories and photographs of their visits were sent to newspapers in their districts. Twice each year he summarized the center's finances to agricultural leaders of the OARDC Support Council. Whenever he could, Payne grabbed a member of the Wooster Area Chamber of Commerce by the elbow to tell him or her how much OARDC had done with so little money. These leaders shared the center's needs with legislators and then-Governor George Voinovich. In budget years, Payne, Smith, and Bobby Moser, Ohio State's vice president for agricultural administration and dean of the College of Food, Agricultural, and Environmental Sciences, teamed up with lay leaders to visit legislators in their districts to tell them what OARDC and OSU Extension were doing for all Ohioans.

These efforts produced results. Between January 1993 and 1999, state appropriations to OARDC increased more than eighty percent. For the 1997 fiscal year, the legislature gave OARDC $26.6 million, the largest amount in the center's history, though appropriations for the 1998 fiscal year were even larger, $30.6 million. State and federal monies for 1997 totaled $43.5 million, which set another record. Through a combination of thrift and increased appropriations, Payne and the center's scientists worked together to retire the deficit by the middle of 1997. OARDC balanced its books for the first time since Kottman had been director. But Payne was quick to point out that this feat would not have been possible without help from Bobby Moser; Keith Smith; associate directors Jim Brown and Skip Nault; Bob Oyer, the center's fiscal manager; and the center's many stakeholders throughout Ohio.

Because Payne drew on the talents of everyone, OARDC has a new vitality. One can hear the click of a computer keyboard in every office. Fax machines churn out documents, and scientists open e-mail that comes from around the globe. For that matter, every business card now has an e-mail address and fax number. The center has an Internet web site with dozens of links that lead one through a maze of research on everything from turkeys to field tomatoes, and

In this publicity photo, Tom Payne and friends—the children of several OARDC employees—model the plastic pig noses to be given away at the first BioHio (1994).

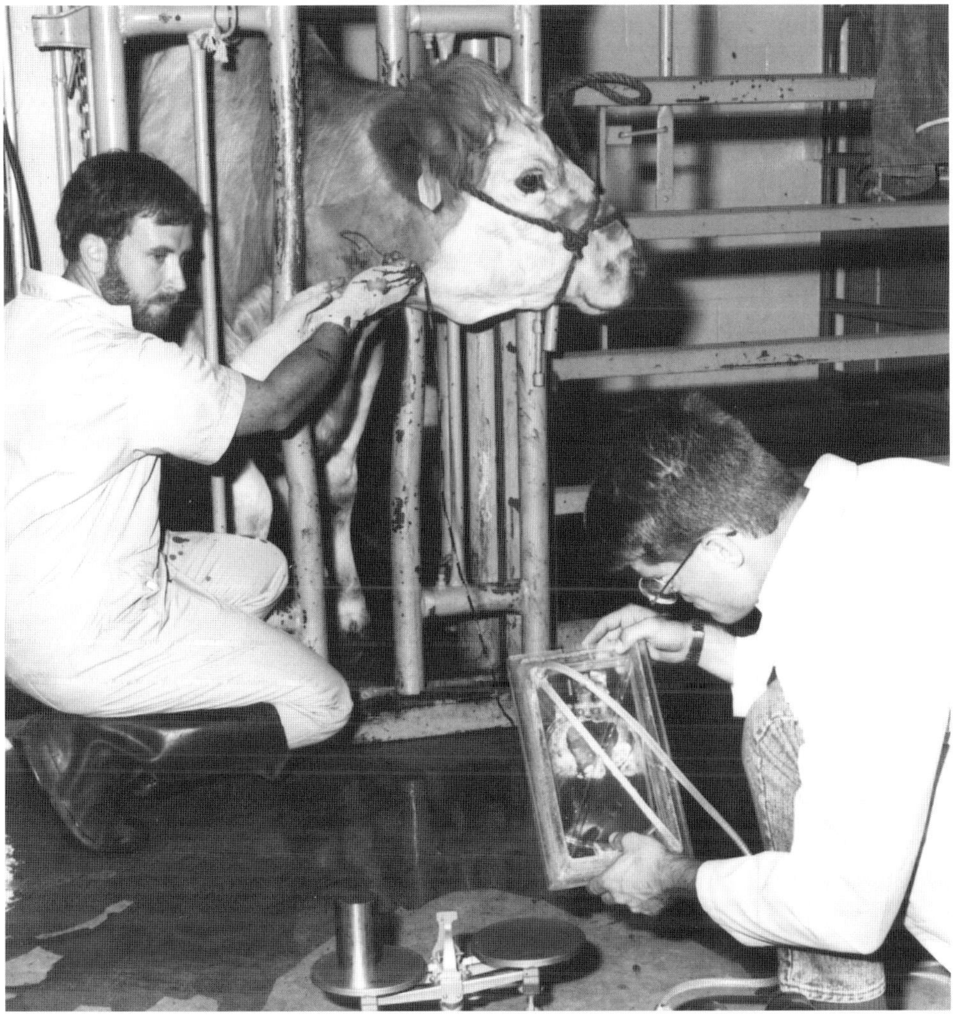

Food Animal Health Research Program personnel draw a blood sample from an OARDC cow in 1994. At the time, the center was cooperating with the Cleveland Clinic on a project: blood from center cattle was being sent to the clinic to aid in the testing of artificial human hearts. The cows were not harmed. The research was one of many projects featured at BioHio in 1994.

into the Molecular and Cellular Imaging Center, the academic departments, the library, the Research Services Building, the Secrest Arboretum, the Laboratory for Pest Control Application Technology, and down many other corridors of agricultural science. New greenhouses have taken shape, their glass and plastic gleaming in the sunlight. The library has new stacks on which stretch volume upon volume of journals. Within their pages are articles by scientists with the center. Thanks to these scientists' efforts, the 1993–1997 report from the Institute for Scientific Information placed Ohio State's output in agricultural research in the top ten of U.S. colleges of agriculture.

Indeed, science is at the heart of OARDC's rejuvenation. Science is part of the Thanksgiving meal that brings out the pilgrim in most Americans. What has science to do with cranberry sauce, sweet potatoes, and turkey? Science has solved a problem that began when breeders developed a turkey whose breast grew so massive that its legs could no longer support such bulk. Deformed or broken legs were the result. The birds grew more slowly, suffered, or died.

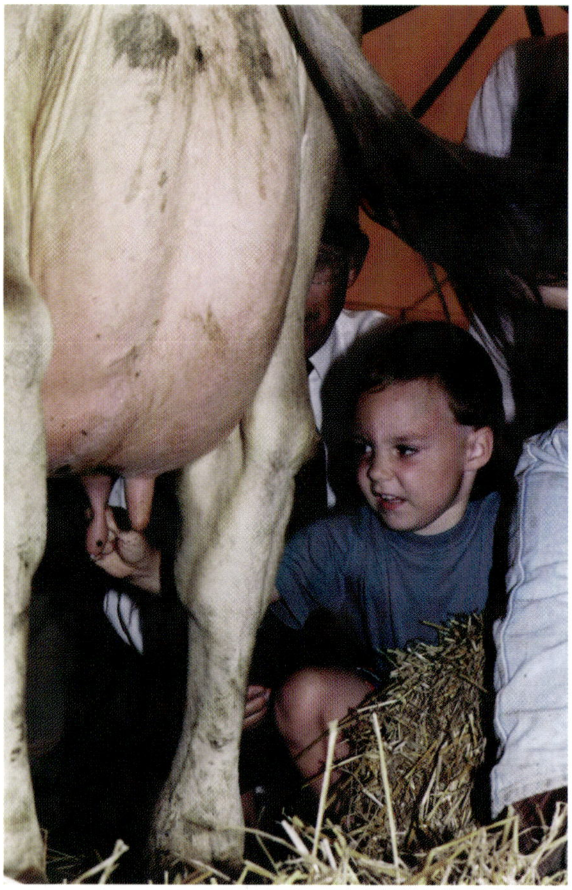

Milking a cow by hand—not a common practice any more—at BioHio 98. The boy is unidentified. Tom Payne is in the background, udderly obscured.

Producers lost income. For Karl E. Nestor, an animal scientist, the cause of this problem was simple—breeders had concentrated on breast girth at the expense of leg strength. To right the balance, Nestor bred turkeys with stout leg bones that could carry the bird's bulk. In the process, Nestor did his part to preserve the feast that stretches back to the first colonists and that today gives millions of people an excuse for watching football on a Thursday in November.

In plant pathology, Laurence V. Madden continued the fight against pathogens that Augustine Selby began more than a century ago. In 1980 Madden received a doctorate in plant pathology from Pennsylvania State University. Three years later he joined the center, where, with the aid of a computer, Madden has, among other things, described how raindrops spread anthracnose fruit rot, a fungal disease, with each splash on a field infested with the fungi. As a result of his achievements, Madden won the Ciba-Geigy award for research from the American Phytopathological Society, the center's Senior Faculty Research Award, and Ohio State's Distinguished Scholar Award. In addition to his research, Madden co-authored the textbook *Introduction to Plant Disease Epidemiology*. In 1992, Madden was elected a fellow of the American Association for the Advancement of Science, and four years later became only the third Ohioan to be president of the American Phytopathological Society.

Traditional crops remain a focus of research, as they were since the center's founding in 1882. Steven K. St. Martin, a soybean breeder, has worked to adapt soybeans to no-till farming. In the absence of plowing, plant residues accumulate and serve as a breeding ground for insects, slugs, and pathogens, and so Martin has bred soybeans with broad resistance to these pests. Likewise, Kim Campbell and Patrick E. Lipps have developed a high-protein oat that is resistant to barley yellow dwarf virus.

Of course, one does not need to tell Frank Hall of the Laboratory for Pest Control Application Technology how troublesome insects can be. But he does not kill them by spraying insecticides with gusto as Joseph Polivka did. Since Rachel Carson published *Silent Spring* in 1962, Americans have demanded safer insecticide use. To satisfy this desire, Hall and colleagues have designed equipment and techniques to reduce the drift of insecticides onto anything other than the insect to be killed.

But OARDC scientists have also devoted themselves to more than helping farmers raise crops and livestock, as they've done since 1882 when William Lazenby took the first tentative step toward a science that went beyond the farm by analyzing drinking water for purity. Nutritionist Jean T. Snook has described how saturated and unsaturated fats affect cholesterol in blood. In the Department of Human Development and Family Science, David W. Andrews has worked on explaining why peers sometimes influence a child to misbehave. And Robert Hughes Jr. has sought to understand how African-American, white, and Hispanic women cope with the stress of divorce and the demands of raising children alone.

This broad mission has included a commitment to the environment that stretches back to the conservation movement of Edmund Secrest's day. Plant pathologists and agricultural engineers have designed ways of composting plant and animal wastes that minimize odor and improve the use of compost in agriculture. Poultry scientists have labored to limit noise and odor from broiler grow-out farms that are located near populated areas. Horticulturists have promoted crop rotation and companion planting as ways for gardeners to

(left) Laurence V. Madden, left, and Michael A. Ellis, both of the Department of Plant Pathology, in a photograph taken for the 1988 OARDC annual report. Madden and Ellis developed a computerized predictor for black rot, a major disease of grapes in Ohio. By using the device, growers could reduce their fungicide sprays from ten or more a season to as few as two. (right) Anne E. Dorrance and Patrick E. Lipps, also of the Department of Plant Pathology, in a photograph in the 1998 annual report. The accompanying story featured research on developing disease-resistant soybeans and on management practices that reduce infections.

David B. Min, Department of Food Science and Technology, left, and Steven K. St. Martin, Department of Horticulture and Crop Science, in Columbus in 1995. St. Martin helped develop two new food-grade soybean varieties, called FG1 and FG2, for Ohio farmers. The varieties were for use in making tofu and for potential sale in Asia. Min adapted a processing method to remove some of tofu's "beany" taste, a way to boost tofu and soy milk sales in the United States.

reduce insect damage while using fewer or no insecticides. Their ideas help suburbanites grow healthier tomatoes, peppers, cabbage, and potatoes in the backyard garden. "Science," Payne remarked, "must help farming to be environmentally sound and socially acceptable."

The quality and breadth of this research underscores that Payne did his job well. Indeed, his success in Ohio made him an attractive leader to experiment stations and land-grant colleges throughout the country. In January 1999, Payne became director of the Missouri Agricultural Experiment Station and dean of the College of Agriculture, Food, and Natural Resources at the University of Missouri–Columbia. As had been the case throughout his life, Payne was eager for new challenges, though he admitted he would carry fond memories of the Ohio center wherever he goes. Payne left having restored morale, balanced the budget, and reinvigorated the center. The OARDC family missed him and would cheer his successes in the "Show Me" State.

The end of Payne's tenure provided an opportunity to survey the sweeping changes of the last 116 years. In 1882, William Lazenby had just three part-time scientists and $3,000. In 1998, Payne could call on the talents of 250 scientists on the Wooster and Columbus campuses and a budget of nearly $50 million. Today OARDC has 14,500 times more money than it had in 1882. Then, it had no land of its own, but occupied seventeen acres on Ohio State's Columbus campus. Today the center spans 7,500 acres in Wooster, at ten outlying branches, and at the Piketon Research and Extension Center, along with extensive facilities in Columbus. Through OSU Extension, the center reaches Ohioans in all eighty-eight counties. In 1882, the station languished in the shadow of Ohio State, and few people in Columbus thought much about agricultural science. Today OARDC is the agricultural research arm of the university. The graduate student who is serious about agricultural science can study in Wooster as well as in Columbus. In the center's early years, Lazenby struggled to gain recognition for the station with Ohioans. Today people in China, India, Mexico, Argentina, Chile, Brazil, Peru, Russia, Ukraine, and Nigeria know of OARDC and have met its scientists. In 1882, Lazenby covered the station's research in a few sentences in the annual report. In 1998, the

OARDC scientists, OSU Extension specialists, and members of Ohio's Amish community teamed up to develop a new slow-moving vehicle sign and new reflective markings. They found materials that, compared to old ones, are visible twice as far and last twice as long. Their work established new national standards.

Ellen I. Hock, Department of Family Relations and Human Development (now Human Development and Family Science), College of Home Economics (now Human Ecology), in 1988. She found that depression is common among career-oriented mothers of young children and is worse for mothers who want to work yet stay home. The study gained national attention.

center's research appeared in leading scientific journals and was summarized at conferences throughout the world. Then, the scientists primarily were local men. Today they are fellows of the American Association for the Advancement of Science and members and officers of the American Phytopathological Society, the Poultry Science Association, the Entomological Society of America, the Society of American Foresters, the American Society of Agronomy, and other esteemed professional organizations. If these activities are not enough, OARDC scientists are also referees and editors for leading journals.

Along with these changes, OARDC retains much of its past. In its early years, Lazenby labored to build broad support for the station. He sought to attract urbanites by analyzing drinking water for purity, by testing varieties of turf grass for suitability to Ohio's soils and climate, and by instructing homeowners on the care of rosebushes. Under Thorne, Edmund Secrest brought the conservation movement to Ohio and created an arboretum that today appeals to plant lovers, gardeners, landscapers, and tourists. Kottman kept the center

Rhododendrons and azaleas, a favorite of visitors, bloom in mid- to late May in Secrest Arboretum. (over) Rhododendrons at the Administration Building at Wooster.

relevant to Ohioans in the city and in the suburbs by promoting the development of trees to shade city streets. Under Payne, OARDC continued to appeal to all Ohioans through BioHio, by reducing pesticide use, and by composting plant and animal wastes. The ecology that kids learn in their schools is being practiced by center scientists.

Since its first days in Columbus, the center has been committed to helping farmers provide all Ohioans with abundant and nutritious food. Thorne "taught" the acre to produce more food; Rummell gave Ohioans three square meals a day; Kottman spoke of his hope that farms would yield seventy bushels per acre of soybeans; and Payne touted the center's importance in providing research that helps agriculture be efficient, environmentally sound, economically viable, and socially acceptable. From the beginning, OARDC has been committed to doing the best science. This devotion links Thorne to Payne, Selby to Madden, Stringfield to St. Martin, Krauss to Nestor.

The future will bring new directors, larger appropriations, and more scientists to Columbus, Wooster, Piketon, and the outlying branches. OARDC will change in ways that no one today can predict, except that it is safe to say that its future will be no less eventful than its past. One should be assured, however, that the center will continue to benefit all Ohioans. The National Archives in Washington, D.C., reminds all Americans that "the past is prologue." May the Ohio Agricultural Research and Development Center's future be as full of achievements as the past and present have been.

INDEX